T0264695

Nanostructured Ceramic Oxides for Supercapacitor Applications

Nanostructured Ceramic Oxides for Supercapacitor Applications

Edited by
Avinash Balakrishnan
K. R. V. Subramanian

CRC Press
Taylor & Francis Group
Boca Raton London New York

CRC Press is an imprint of the
Taylor & Francis Group, an **informa** business

CRC Press
Taylor & Francis Group
6000 Broken Sound Parkway NW, Suite 300
Boca Raton, FL 33487-2742

First issued in paperback 2017

© 2014 by Taylor & Francis Group, LLC
CRC Press is an imprint of Taylor & Francis Group, an Informa business

No claim to original U.S. Government works

ISBN-13: 978-1-4665-7690-2 (hbk)
ISBN-13: 978-1-138-07267-1 (pbk)

This book contains information obtained from authentic and highly regarded sources. Reasonable efforts have been made to publish reliable data and information, but the author and publisher cannot assume responsibility for the validity of all materials or the consequences of their use. The authors and publishers have attempted to trace the copyright holders of all material reproduced in this publication and apologize to copyright holders if permission to publish in this form has not been obtained. If any copyright material has not been acknowledged please write and let us know so we may rectify in any future reprint.

Except as permitted under U.S. Copyright Law, no part of this book may be reprinted, reproduced, transmitted, or utilized in any form by any electronic, mechanical, or other means, now known or hereafter invented, including photocopying, microfilming, and recording, or in any information storage or retrieval system, without written permission from the publishers.

For permission to photocopy or use material electronically from this work, please access www.copyright.com (http://www.copyright.com/) or contact the Copyright Clearance Center, Inc. (CCC), 222 Rosewood Drive, Danvers, MA 01923, 978-750-8400. CCC is a not-for-profit organization that provides licenses and registration for a variety of users. For organizations that have been granted a photocopy license by the CCC, a separate system of payment has been arranged.

Trademark Notice: Product or corporate names may be trademarks or registered trademarks, and are used only for identification and explanation without intent to infringe.

Library of Congress Cataloging-in-Publication Data

Nanostructured ceramic oxides for supercapacitor applications / edited by Avinash Balakrishnan and K.R.V. Subramanian.
   pages cm
  Includes bibliographical references and index.
  ISBN 978-1-4665-7690-2 (hardback)
  1. Supercapacitors--Materials. 2. Oxide ceramics. 3. Nanostructured materials. I. Balakrishnan, Avinash. II. Subramanian, K. R. V.

TK7872.C65N36 2014
621.31'5--dc23
                             2013036693

Visit the Taylor & Francis Web site at
http://www.taylorandfrancis.com

and the CRC Press Web site at
http://www.crcpress.com

Contents

Preface

The field of high-energy and power storage devices, the domain of super-capacitors (or ultracapacitors as they are known), has seen a recent upsurge in market development and usage, research and development activity, and significant and rapid strides in device development. Thus, we have foreseen a renewed interest in this emerging field (and specifically the field of ceramic oxide–based supercapacitors) by the student population and well-educated next generation of scientists and engineers. This textbook originated from our own intense and sustained research in the field of supercapacitors and AB's substantial research background in the area of ceramic materials and their application to supercapacitor development. This book is designed to cater to a broad base of seniors and graduate students having varied backgrounds such as physics, electrical and computer engineering, chemistry, mechanical engineering, materials science, nanotechnology and even to a reasonably well-educated layman interested in state-of-the-art devices. Given the present unavailability of a "mature" textbook having a suitable breadth of coverage (although basic books and a plethora of journal articles are available with the added difficulty of referring to multiple sources), we have carefully designed the book layout and contents with contributions from well-established experts in their respective fields. This book is aimed at undergraduate seniors and graduate students in all of the disciplines mentioned earlier.

The textbook consists of six well-rounded chapters arranged in a logical, distilled manner. Each chapter is intended to provide an overview, not a review, of a given field with examples chosen primarily for their educational purpose. The student is encouraged to expand on the topics discussed in the book by reading the exhaustive references provided at the end of each chapter. The chapters have also been written in a manner that fits the background of different science and engineering disciplines. Therefore, the subjects have been given primarily a qualitative structure, and in some cases, providing a detailed mathematical analysis.

Each chapter contains several simple, well-illustrated equations and schematic diagrams to augment the research topics and help the reader grasp the fundamental nuances of the subject properly. The progression of chapters is designed in such a way that all background theories and techniques are introduced early on, leading to the evolution of the field of nanostructured ceramic oxide–based supercapacitors. The readers will find this logical evolution highly appealing as it introduces a didactic element to the reading of the textbook apart from the joy of grasping the essentials of an important subject.

We, the editors (KRVS and AB) first and foremost thank the dedicated and respected individual scientists who have written the individual chapters. Their enthusiasm in writing chapters of high quality and delivering them on time after incorporating the review suggestions made the release of the textbook a simplified task for us. In addition, several outstanding graduate students of the Nanosolar and Energy Storage Technologies (NEST), Amrita Centre for Nanosciences, Kochi, Kerala, India, have also made important contributions. We thank Ranjusha, Anjali, Praveen P., Roshny, Lakshmi V., Pratibha, and Vani, in particular, and the other students of the division for their assistance and also Dr. Sivakumar for his assistance. We also thank our Division Director Dr. Shanti Nair and the management of Amrita Institute of Medical Sciences, Kochi, for their support in undertaking this project.

Editors

Dr. Avinash Balakrishnan joined as faculty at Nanosolar Division in 2010, following a postdoctorate at Grenoble Institute of Technology (Grenoble-INP) France. He completed his PhD and MS in Materials Engineering at Paichai University, South Korea. He is an alumnus of National Institute of Technology, Karnataka, where he completed his bachelor's in metallurgical engineering. Dr. Balakrishnan has received several prestigious fellowships and travel grants throughout his academic career that includes a KOSEF fellowship to pursue his master's (2003), Korea Research Foundation Fellowship for his PhD (2005), and CEFIPRA postdoctoral fellowship (2008). His past professional experience also includes being a research scientist at Korea Research Institute of Standards and Science, South Korea, where he extensively worked on high-temperature and high-strength ceramic materials for space, armor, and dental applications. During his postdoctoral research, he mainly focused on simulations and mechanics related to ceramic nanoparticles.

His current research interest includes developing nanostructured ceramic-based oxides as electrode materials for energy storage devices. The concepts being investigated are growth behavior, morphological effect on the charge–discharge mechanism, and effect of secondary particle phase in the electrode to name a few. He is also actively leading five government-funded projects related to high–energy density supercapacitors. He has coauthored more than 70 research publications in reputed international journals such as *Acta Materialia, Journal of American Ceramic Society*, and *Journal of Materials Research*. He has filed 8 Indian patents in the area of energy storage. He is currently the associate editor of the *Journal of Nanoscience and Nanotechnology*.

Dr. K.R.V. Subramanian is currently an associate professor with Amrita Centre for Nanosciences, Kochi, India. He joined Amrita in July 2008. He obtained his PhD from Cambridge University, England, in 2006. His doctoral dissertation focused on the development and characterization of several unary and multicomponent oxide electron beam resists that are suitable for high-resolution electron beam nanolithography. His doctoral work was highly commended and resulted in Engineering and Physical Sciences Research Council (EPSRC) proposals and publications in high-impact journals, such as *Nanoletters* and *Advanced Materials*. He is a fellow of the Cambridge Commonwealth Society. His previous alma mater for his bachelor's and master's degrees in engineering include the National Institute of Technology, Tiruchirapalli, and the Indian Institute of Science (IISc), Bangalore. He also worked as a postdoctoral research associate at the University of Illinois at Urbana–Champaign, where he was involved in device fabrication (cancer cell charge sensing) and high-resolution electron beam lithography studies on a widely used electron beam resist poly(methyl methacrylate). He continued his postdoctoral research at IISc, Bangalore, where he oversaw work on zinc oxide and conducting polymers. Dr. Subramanian has 60 journal publications and 29 conference presentations to his credit. He is also a reviewer for many prestigious journals such as *Energy and Environmental Science, Nanoscale, Journal of Physical Chemistry, Chemical Communications*, and *RSC Advances*. Dr. Subramanian's research interests encompass nanostructured photovoltaics, materials for energy storage, supercapacitors, batteries, nanofabrication, device processing, functional nanomaterials, and nanomaterials for cancer therapy. He was the co-principal investigator for large government funded projects in the area of photovoltaics and energy storage.

Contributors

Heather Andreas received her PhD degree in electrochemistry from the University of Calgary, Canada, in 2004. She pursued her postdoctoral studies at the University of Ottawa with an emphasis on electrolyte-related issues of supercapacitors. In 2006, she joined as assistant professor at Dalhousie University in Halifax, where she received the university faculty award from the Natural Sciences and Engineering Research Council and the Leaders Opportunity Award from the Canadian Foundation for Innovation. In July 2011, she accepted a position as associate professor at Dalhousie. Her research focuses on the use of electrochemical methods to study and optimize alternative energy storage with an emphasis on supercapacitors.

S. Jayalekshmi is a professor in the Department of Physics of the Cochin University of Science and Technology, Kerala, India. Her research areas include polymer light-emitting diodes, rechargeable lithium-ion cells, nonlinear optics, nanophotonics, and bio-nanophotonics. Eleven students have completed PhDs under her supervision and eight more are currently working for PhDs under her supervision. She has contributed more than 60 publications in international journals. She is a member of the Society of Photo-Optical Instrumentation Engineers (SPIE) and the Electrochemical Society (ECS) and a life member of the Material Research Society of India. Her research laboratory, "Division for Research in Advanced Materials," (DREAM) can be accessed at http://physics.cusat.ac.in/research/dream_lab/index.html.

Meisam Valizadeh Kiamahalleh received his BSc in chemical engineering from University of Mazandaran, Iran, in 2006, where he proceeded with his first research project on "reverse osmosis," under the guidance of Prof. Ghasem Najafpour. In 2007, he moved to Malaysia to pursue his MSc in chemical engineering in Universiti Sains Malaysia on carbon nanotube-based nanocomposites for the supercapacitor electrode materials under the supervision of S. H. S. Zein. He has six ISI journal articles and four international conference papers to his credit. Meisam is now pursuing his PhD on novel graphene-based nanoporous materials for controlled drug delivery under the supervision of Prof. Mark Biggs at the School of Chemical Engineering, University of Adelaide, Australia.

Praveen Pattathil graduated from Bharathiar University, Tamil Nadu, India, with a master of science degree in physics in 2007. He completed his MTech in nanotechnology at Amrita Nanosolar and Energy Storage Division,

Amrita University, Kochi, India. His research focuses on metal oxide–based supercapacitor/battery material. He has also received a Ministry of New and Renewable Energy (MNRE) fellowship and has more than ten publications in international journals to his credit.

Anand Puthirath graduated from Cochin University of Science and Technology with an MSc in physics in 2010. Currently, he is a PhD student in the Department of Physics, Cochin University of Science and Technology, under the guidance of Dr. S. Jayalekshmi. His research is focused on polymer-based rechargeable battery materials, organic photovoltaic materials, and nonlinear optics active photonic materials. He is a student member of the reputed research organizations SPIE and ECS. He has won the Best Student Paper Award at the International Conference OPAP-2013 held in Singapore, organized by Global Science & Technology Forum (GSTF). He has six publications in international journals and has attended many international and national conferences.

P. Ragupathy is a quick hire scientist in Electrochemical Power Systems Division at the Central Electrochemical Research Institute, Karaikudi. He received his BS, MS, and MPhil degrees in chemistry from St. Joseph's College, affiliated with Bharathidasan University, Tiruchirappalli, India. He obtained his PhD degree in materials electrochemistry from the Indian Institute of Science, Bangalore, India. Before becoming a research fellow at Nanyang Technological University and NUS Nanoscience and Nanotechnology Initiative at Singapore, he was a postdoctoral fellow at the University of Texas, Austin. His research interests include development of new synthesis methods and nanostructured materials for lithium-ion batteries, redox flow batteries, and supercapacitors.

Jaya T. Varkey graduated in applied chemistry from Cochin University of Science and Technology, Kerala. She received her PhD in chemistry from the School of Chemical Sciences, Mahatma Gandhi University, Kerala, India. She developed a novel polymeric resin for solid-phase peptide synthesis during her PhD work. A postdoctoral fellow from the University of Minnesota, USA (1999–2003), she has presented papers in several national and international conferences including the 17th American Peptide Symposium, San Diego. She has won the National Young Scientist award winner in chemistry in 1995. She has to her credit several international papers and a book on peptide synthesis by Lambert Academic Publishing, Germany. She is currently an assistant professor in chemistry at St. Teresa's College, Ernakulam, Kerala, India.

S. H. S. Zein received his PhD in chemical engineering from Universiti Sains Malaysia (USM), where he then joined this department as a lecturer and was promoted to associate professor. His research interests include chemical reaction engineering, catalyst synthesis and characterization, and carbon nanotubes. His current interest is catalysis of plant fibers toward the production of nano-carbonaceous materials and an alternative source of energy, and the development of bone replacement materials. His outstanding academic contribution is also evidenced by the 130 scientific articles he has published over the past 10 years; some of them were listed as the top 25 hottest articles published for a period of time. Apart from authorship, he worked as international grant assessor in many reputed regional and international journals. In the past 10 years, he has been awarded nationally and internationally for his research and teaching. Receiving Excellent Teaching Awards from USM and two awards from the British Inventors Society in 2009 and 2011 are the most recent ones.

1

Capacitor to Supercapacitor: An Introduction

Praveen Pattathil, Nagarajan Sivakumar, and Theresa Sebastian Sonia

CONTENTS

1.1 Historical Background

Supercapacitors, also known as electrochemical capacitors or ultracapacitors, store electrical charge in the electric double layer at an electrode–electrolyte interface, primarily in a high–surface area electrode. The high surface area and the small thickness of the double layer (in the order of angstrom) enable these devices to exhibit high specific and volumetric energy and power density with an essentially unlimited charge–discharge cycle life. The operational voltage in these devices is governed by the breakdown potential of the electrolyte and ranges usually between <1 and <3 V for aqueous and organic electrolytes, respectively.[1–3]

The concept behind the supercapacitor mechanism was known since the late 1800s. The first supercapacitor device, based on double-layer charge storage (using carbon electrodes), was built in 1957 by H.I. Becker of General Electric for which he was granted a U.S. patent (No. 2,800,616).[4] Figure 1.1 shows a diagrammatic representation of this device.

The limitation of Becker's device was that its setup was similar to a battery where both anode and cathode need to be immersed in a container of electrolyte. This dramatically reduced the volumetric energy and power density of the device. The device was never commercialized. However, over the last few decades, supercapacitor technology has grown into a full-fledged industry with sales amounting to several hundred million dollars annually. The unique properties of supercapacitors have often complemented the limitations of batteries and fuel cells. With increasing oil prices, this is an industry that has rapidly expanded in power quality and emerging automobile applications. Earlier supercapacitors could deliver few volts and had capacitance values measuring from microfarads up to several tens of farads. With the latest advancements, cells exhibiting millifarad storage capacity

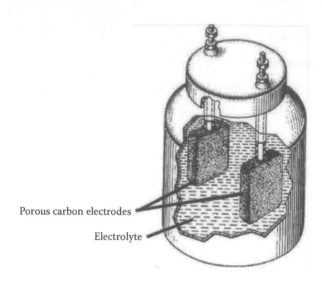

FIGURE 1.1
The General Electric–patented capacitor. (From Becker H.I., Low-voltage electrolytic capacitor. U.S. Patent 2,800,616.)

with exceptional pulse power performance to devices having hundreds of thousands of farads of storage capability and operational voltages up to 1500 V are available in the market. This chapter aims at briefly presenting the historical background of the evolution of supercapacitor technology.

In 1962, Robert A. Rightmire, a chemist at the Standard Oil Company of Ohio (SOHIO), filed a patent application (U.S. patent 3,288,641 awarded in 1966)[5] for a supercapacitor device that he had invented in the format in which it is presently used. The carbon double-layer capacitors were originally developed by SOHIO as a by-product of fuel cell–related development activities. The patent by Robert Rightmire was subsequently followed by another U.S. patent application (No. 3,536,963)[6] by a fellow SOHIO researcher Donald L. Boos in 1970, which laid the foundation for the hundreds of patents and journal articles covering all aspects of supercapacitor technology. At an international conference on double-layer capacitors and similar energy storage devices in 1991, Donald Boos described in detail the phenomenon and his activities in developing prototypes of what he termed "Electrokinetic capacitors." The motivation for this research was initiated by the problem associated with obtaining stable electrical readings on fuel cells. The duration of the stability was attributed to the electrodes being used in the device. This triggered the concept of employing high–surface area carbon electrodes.

However in 1971, SOHIO went into financial crisis, as its Alaska oil pipeline venture took a heavy beating. This led SOHIO to cutback many of its long-term research investments, including the capacitor project. This prompted them to license their supercapacitor prototype for commercialization, but the licensee failed.

The prototypes of the supercapacitors, which SOHIO sent across the globe in its quest to commercialize, harnessed a lot of interest. In 1974, USSR Academy of Sciences reported a paper entitled, "Anomalous Electrical Capacity and Experimental Models of Hyperconductivity" regarding a molecular capacitor. Although this article was received with high skepticism, the research grants were funded, albeit with close inspection.

By this time, that is, in 1975, a second licensee, the Nippon Electric Company (NEC) of Japan, entered into an agreement with SOHIO for this technology. As it started recovering from its financial crunch with the oil production happening and flowing through its Alaska pipeline in 1977, SOHIO also resumed its supercapacitor research activities. By 1978, NEC had invested a lot of time and money in fundamental investigations, expedited manufacturing and marketing capability for its supercapacitor devices, creating large supply chains. Two years later in 1980, the SOHIO introduced its "Maxcap" double-layer capacitor product, having an aqueous electrolyte with capacitance rated up to several farads and exhibiting an operating voltage of 5.5 V. But constant transferring of ownership, to a company called Carborundum, then to Cesiwid, and finally to Kanthal Globar, allowed NEC to take the command as the world leader in this technology with a rich history of developing and marketing supercapacitor products. Hence, NEC is often credited for the term "Supercapacitor" for its electrochemical capacitor products. By 1982, different variants of NEC's supercapacitors were pitched into the market to meet new application requirements.

Beginning in 1985, research efforts were diverted on the utilization of supercapacitors in automobile applications with emphasis on starters for internal combustion engines. The Cold War and the Space Race triggered the Russians to also participate and contribute in this area. A Russian company called MP Pulsar (later changed its name to ELIT) started developing state-of-the-art supercapacitors in 1988 at the Accumulator Plant in Kursk. The following year (1989), ELIT developed the first asymmetric supercapacitors based on the use of a nickel oxyhydroxide as cathode, potassium hydroxide as electrolyte, and activated carbon as anode. These devices were used to power wheelchairs and toys. In 1990, ELIT started focusing on symmetric designs with potassium hydroxide electrolyte and two activated carbon electrodes.

Inspired by these developments, a company by the name The Econd Corporation was founded in 1991 in Moscow, which demonstrated that their product could be used for starters in trucks as well as for hybrid electric vehicles (HEVs). Again in the same year, NEC reported the development of high-performance double-layer capacitors using activated carbon electrodes.

These developments made Russians to invest more into this technology, which led to the opening of a manufacturing unit under a joint stock company called ESMA in Troitsk, Russia, in 1993. Soon, ESMA pioneered the development of devices in a range of sizes from 20 kJ, 14 V modules to 30 MJ 190 V modules used to power heavy-duty automobiles like trucks and buses. These modules exhibited capacitance in the range of 3000–100,000 F per cell.

NEC Japan's innovative idea of using activated carbon electrodes was employed by ESMA for constructing anodes for cells similar to nickel–cadmium (Ni-Cd) batteries used in aircrafts. These ESMA cells offered very good power and energy storage performance with high cycling stability at a reasonable cost.

The market potential of this technology grabbed worldwide attention. In 1994, Commonwealth Scientific and Industrial Research Organization of Australia joined hands with Plessey Ducon, a manufacturer of passive components. Two years later, the research and development team produced a carbon and organic electrolyte–based spiral-wound supercapacitor with an energy density of 9 Wh/kg and time constants ranging from seconds to milliseconds.

In 1996, with the rising competition worldwide, NEC introduced several new products to the market, including the FG series, the high-power FT series, and the FC series under the brand name "Tokin."

In 1997, supercapacitors made history when it was first disclosed that such systems can be designed to deliver energy densities exceeding 10 Wh/kg. These capacitor banks opened up avenues for propulsion applications in heavy-duty vehicles, like trucks and buses, where it could store about 30 MJ of energy at 190 V. Dr. Arkadiy Klementov of ESMA delivered a disclosure presentation titled Application of Ultracapacitors as Traction Energy Sources at the 7th International Seminar on Double-Layer Capacitor and Similar Energy Storage Devices. ESMA design was an asymmetric capacitor (U.S. patent no. 6,222,723),[7] which essentially used one battery electrode coupled with a carbon electrode. The combination offered advantages in terms of higher specific energy, higher operating voltage, and low cost.

By this time, Nippon Chemi-Con (NCC) of Japan, a billion-dollar manufacturing company which had earned the distinction of being the largest producer of aluminum electrolytic capacitors, entered the market with their supercapacitors under the brand name "DLCAP." NCC started its mass production in 1998 with both spiral-wound and prismatic cell designs having 3000 F and 2.5 V ratings. NCC used a propylene carbonate solvent as its electrolyte, which does not have the fire and health issues associated with some of the alternative organic electrolytes. They employed a resealable valve approach, which was a boost to capacitor production.

During the same year, that is 1998, the Daewoo group of South Korea designed its first supercapacitor product and named it "NessCap." With both public and private investors pitching in funds, NessCap started capturing the market with its broad product line of supercapacitor variants. The design of NessCap products was based on an organic electrolyte with a spiral-wound prismatic cell configuration showing capacitance from a few farads to 5000 F and was rated at 2.7 V, one among the highest voltages in the industry.

However, Japanese domination in the market still remained. The year 1999 saw Panasonic introducing its high-performance "UpCap" supercapacitor, rated at 2000 F and 2.3 V, for transportation applications including HEVs.

It offered a waterproof construction at a low cost. The supersealed package also ensured that no heat was entrapped during the operation. This feature was important, because heat was generated during the charging/discharging of hybrid vehicles (i.e., during regenerative braking and acceleration) and had to be dissipated.

Supercapacitors inherited new engineering designs over a period of time with new applications. For instance, in 2001 a supercapacitor design exhibiting spiral-wound series and a low-profile thin-format series, both having an organic electrolyte, appeared in the market. This was introduced by "Tokin" of Japan, which later changed its brand name to "NEC-Tokin" in 2002.

The year 2002 saw a lot of developments in this area. At the centennial meeting of the Electrochemical Society, Asahi Glass's Dr. T. Morimoto described an asymmetric supercapacitor design in which both electrodes were carbonaceous in nature and used an organic electrolyte containing a lithium salt. These prototypes showed a volumetric energy density of 16 Wh/L. Today several companies are following this approach because of the higher operating voltage (i.e., 4.2 V) and higher energy density the design has offered.

The same year saw Maxwell Technologies acquiring supercapacitor manufacturer Montena Components of Rossens, Switzerland. This acquisition resulted in a lot of new designs coming into the market. For example, metalized carbon cloth electrode material was replaced by carbon-coated aluminum foil current collectors, which were then made into cylindrical cells. Maxwell also launched its broad range of supercapacitor products called "BoostCaps." Currently, Maxwell is manufacturing larger devices that are being used in many applications, including HEVs, power generation systems like wind turbines, and internal combustion engine starters. Maxwell licensed its technology to EPCOS in Germany; however, EPCOS elected to exit this market in late 2006, as it was unsure about the time it would require to achieve the break-even point of its investment. However, this did not stop Maxwell Technologies from becoming the leading U.S. producer of supercapacitors.

In 2007, ESMA and its partner American Electric Power came out with a significantly cheaper lead oxide/sulfuric acid/activated carbon asymmetric supercapacitor which could store large amounts of electricity from the utility grid at night. This storage system was designed to last for ten years and exhibited good cycle numbers. Such asymmetric systems could offer tough competition for many existing battery systems. It should be noted that ESMA had reported good cycle efficiency, with an estimated cycle life greater than 5000 cycles. This system had energy density much higher than that of symmetric supercapacitors but less than that of lead-acid batteries. Figure 1.2 provides the important phases in supercapacitor development.

Today, the largest market for supercapacitors is in uninterruptible power supplies and power quality applications.[8] Since 80% of power line disturbances last less than a second, supercapacitors can provide a better solution than batteries because they offer the lowest cost per unit power and

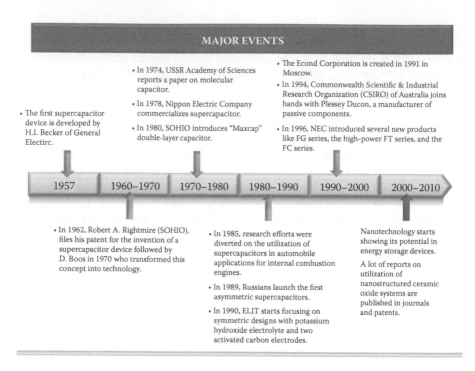

FIGURE 1.2
Timeline showing the important phases in supercapacitor development.

have the longest operational life. Another product with a quickly growing market is HEVs.[9] Over 50 state-of-the-art HEV models are expected to be on the road by 2015, an outstanding 300% growth rate that is expected to last at least a half decade. Cell phones and digital cameras will also be another major application area. It is projected that a compounded annual growth rate for a supercapacitor is about 27%.

With the onset of nanotechnology, the uses of nanostructured electrodes are also expanding. This enhances the possibilities and practicality of supercapacitors for heavy-duty applications. With huge advancements occurring in nanotechnology, more efficient supercapacitors have been reported. The knowledge of these changes also broadens the possibilities for future advance storage devices. Nanostructured metal oxides present an attractive alternative as an electrode material because of their high specific capacitance and low resistance, making it easier to design high-energy and power density supercapacitors. Extensive research into ruthenium oxide (RuO_2) has been conducted for military applications, where cost is less of an issue than it is for commercial ventures.[10–12] Academic researchers have focused on other cheaper metal oxides as replacement for RuO_2, but the selection has been limited by the usage of concentrated H_2SO_4 as an electrolyte. It was hypothesized that high capacitance and fast charging were largely a result of hydrogen adsorption, so a strong acid was, therefore, necessary to provide good proton conductivity. This limited the range of possible electrode materials, as most metal oxides dissolute quickly in acidic solutions. Milder aqueous solutions, such as KCl, have therefore been considered for use with metal oxides such as MnO_2.[13–15] Although many of these ceramic oxides currently appear to possess lower specific capacitances than ruthenium oxides, the lower cost and milder electrolyte may be enough of an advantage to make them a suitable candidate.

Table 1.1 provides a summary of the worldwide research activities and development on supercapacitors.

TABLE 1.1

Worldwide Research Activities and Development on Supercapacitors

Country/Company (Product)	Description of the Technology	Device Characteristics (Voltage, Capacitance, Energy Density, and Power Density)	Status/Availability
Australia – Cap XX (Carbon particulate composites)	Spiral wound and monoblock, particulate with binder, organic electrolyte	3 V, 120 F, 6 Wh/kg, 300 W/kg	Packaged prototypes
France – aft (Carbon particulate composites)	Spiral wound, particulate with binder, organic electrolyte	3 V, 130 F, 3 Wh/kg, 500 W/kg	Packaged prototypes
Japan – NEC (Carbon particulate composites)	Monoblock, multicell, particulate with binder, aqueous electrolyte	5–11 V, 1–2 F, 0.5 Wh/kg, 5–10 W/kg	Commercial
Japan – Panasonic (Carbon particulate composites)	Spiral wound, particulate with binder, organic electrolyte	3 V, 800–2000 F, 3–4 Wh/kg, 200–400 W/kg	Commercial
Russia – ESMA (Hybrid)	Double-layer/ Faradaic, monoblock multicell modules, carbon/nickel oxide/KOH	1.7 V cells, 50,000 F, 8–10 Wh/kg, 80–100 W/kg	Commercial
Russia – ELIT (Carbon particulate composites)	Bipolar, multicell carbon with H_2SO_4	450 V, 0.5 F, 1 Wh/kg, 900–1000 > 100,000 cycles	Commercial
Sweden – Superfarad (Carbon-fiber composites)	Monoblock, multicell carbon cloth on aluminum foil, organic electrolyte	40 V, 250 F 5 Wh/kg, 200–300 W/kg	Packaged prototypes

(Continued)

TABLE 1.1 (*Continued*)

Worldwide Research Activities and Development on Supercapacitors

Country/Company (Product)	Description of the Technology	Device Characteristics (Voltage, Capacitance, Energy Density, and Power Density)	Status/Availability
U.S. – Evans (Hybrid)	Double-layer/ electrolytic, single cell, monoblock, RuO_2, tantalum powder dielectric, H_2SO_4	28 V, 0.02 F 0.1 Wh/kg, 30,000 W/kg	Packaged prototype
U.S. – Los Alamos National Lab (Conducting polymer films)	Single-cell, conducting polymer PFPT. On carbon paper, organic electrolyte	2.8 V, 0.8 F 1.2 Wh/kg, 2000 W/kg.	Laboratory prototype
U.S. – Pinnacle Research Institute (Mixed metal oxides)	Bipolar, multicell, RuO_2, on titanium foil, H_2SO_4	15 V, 125 F & 100 V, 1 F 0.5–0.6 Wh/kg, 200 W/kg	Packaged prototypes
U.S. – US Army, Fort Monmouth (Mixed metal oxides)	Hydrous RuO_2, bipolar, multicell, H_2SO_4	5 V, 1 F 1.5 Wh/kg, 4000 W/kg	Unpacked lab prototype
U.S. – PowerStor (Aerogel carbons)	Spiral wound, aerogel carbon with binder, organic electrolyte	3 V, 7.5 F 0.4 Wh/kg, 250 W/kg	Commercial
U.S. – Maxwell (Carbon fiber composites)	Monoblock, carbon cloth on aluminum foil, organic electrolyte	3 V, 1000–2700 F 3–5 Wh/kg, 400–600 W/kg	Commercial

Source: Burke A., *J Power Sources*, 91, 37–50, 2000.

Abbreviations: PFPT, poly (3-(4-fluorophenyl)thiophene); RuO_2, ruthenium oxide; Wh, watt-hour.

1.2   Scope of the Book

As the title suggests, the focus of this work is on supercapacitors based on nanostructured ceramic oxides. Ceramic oxides have been selected due to their excellent stability in different electrochemical systems. The justification for the usage of ceramic oxides as electrode materials for supercapacitors is their potential to attain high capacitance at a low cost. This book covers in detail the fundamental science, principles, and measurement techniques associated with the performance evaluation of supercapacitors (Chapters 2–4). The heart of this book (Chapters 5–6) aims to give a self-contained account on the development of metal oxide–based electrodes for supercapacitor applications. The final chapter concludes the discussion by pointing out the future scope and directions of nanotechnology in creating next-generation supercapacitors.

References

1. Burke A. Ultracapacitors: Why, how, and where is the technology? *J Power Sources* 2000; 91: 37–50.
2. Chu A, Braatz P. Comparison of commercial supercapacitors and high-power lithium-ion batteries for power-assist applications in hybrid electric vehicles: I. Initial characterization. *J Power Sources* 2002; 112: 236–246.
3. Conway BE. Transition from "supercapacitor" to "battery" behavior in electrochemical energy storage. *J Electrochem Soc* 1991; 138: 1539–1548.
4. Becker HI. Low voltage electrolytic capacitor. U.S. Patent 2800616.
5. Rightmire RA. Electrical energy storage apparatus. U.S. Patent 3288641; 1966.
6. Boos DL. Electrolytic capacitor having carbon paste electrodes. U.S. Patent 3536963; 1970.
7. Razoumov S. Asymmetric electrochemical capacitor and method of making. U.S. Patent 6222723; 2001.
8. Available from: http://scitizen.com/nanoscience/supercapacitors-start-a-revolution-inenergy-storage-devices_a-5-174.html.
9. Available from: http://www.nanotechwire.com/news.asp?nid = 8048.
10. Miller JP, Dunn B, Tran TD, Pekala RW. Deposition of ruthenium nanoparticles on carbon aerogels for high energy density supercapacitor electrodes. *J Electrochem Soc* 1997; 144: L309–L311.
11. Bullard GL, Sierra-Alcazar HB, Lee HL, Morris JL. Operating principles of the ultracapacitor. *IEEE Trans Magn* 1989; 25: 1.
12. Jayalakshmi M, Balasubramanian K. Simple capacitors to supercapacitors: An overview. *Int J Electrochem Sci* 2008; 3: 1196–1217.
13. Ahn YR, Song MY, Jo SM, Park CR. Electrochemical capacitors based on electrodeposited ruthenium oxide on nanofibre substrates. *Nanotechnology* 2006; 17: 2865–2869.
14. Ma S, Nam K, Yoon W, Yang X, Ahn K, Oh K, Kim K. A novel concept of hybrid capacitor based on manganese oxide materials. *Electrochem Commun* 2007; 9: 2807–2811.
15. Jiang JH, Kucernak A. Electrochemical supercapacitor material based on manganese oxide: Preparation and characterization. *Electrochim Acta* 2002; 27: 2381–2386.

2

Electrochemical Cell and Thermodynamics

Jaya T. Varkey, P. Anjali, and Vara Lakshmi Menon

CONTENTS

2.1   Introduction to Electrochemical Systems

Electrochemistry represents a dynamic field with many applications in today's industrial economy. Electrochemical energy systems are devices that directly convert chemical energy into electrical energy. Fuel cells like batteries or galvanic cells are the most important examples of electrochemical systems. The understanding and design of electrochemical systems are both fascinating and challenging because of the simultaneous collaboration of mass transport, charge transport, electrochemical kinetics, and thermal effects. Electrochemical processes have important applications in both nature and industry. This includes generation of chemical energy in photosynthesis, electroplating, batteries, supercapacitors, and also in the biomedical field.

Electrochemistry explains the relationship between chemical changes and resulting electricity. It deals with the conversion of chemically stored energy directly into electrical work. This includes the study of chemical changes caused by passing an electric current through a medium, as well as the generation of electric energy by chemical reactions. It also connects the study of electrolyte solutions and their chemical equilibrium.[1-3] Chemical reactions, which require energy, can be carried out at the surfaces of electrodes in cells connected to external power supplies. From such chemical reactions, one can learn about the nature and properties of the chemical species contained in the cells, and this knowledge can also be used for the preparation of new chemicals. Electrochemical cells that produce electric energy from chemical energy form the basis of batteries and the fuel cells.

2.2   Electrochemical Cell

The components of an electrochemical cell are two half-cells, made up of an electrode kept in contact with an electrolyte. The electrode can be a conductor (such as a metal or carbon) or a semiconductor. The electrolyte is a solution consisting of charge-carrying ions. For example, a solution of sodium chloride ($NaCl$) in water is an electrolyte containing sodium ions and chloride ions. An electric field applied across this solution causes the ions to move toward oppositely charged fields, Na^+ toward the negative and Cl^- toward the positive field.[4] This movement and the subsequent buildup of charges create a potential difference. The movement of the electrons brought about by the chemical reactions causes a current to flow through the electrodes. In all the electrochemical cells, it is the oxidation-reduction (redox) reaction which leads to the development of potential. Oxidation and reduction reactions taking place simultaneously is called a redox reaction. The system in which such redox reactions take place to produce electrical energy is called an electrochemical cell, which is often called a galvanic cell.[4-7]

The half-cells are connected by a separator that permits the movement of charged ions between the half-cells but avoids the mixing of the electrolytes. The separator can consist of a salt bridge, or it can be an ion exchange membrane or a sintered-glass disk. Both the half-cells can be kept in contact with the same electrolyte as well. In galvanic cells, spontaneous reactions occur at the interface between the electrode and the electrolyte, when the two electrodes are connected. In electrolytic cells, reactions are made to occur at these interfaces with the help of an external power supply connected to both electrodes. The power supply provides electrical energy, which is converted to chemical energy in the form of the products of the electrode reactions.

A typical galvanic cell is shown in Figure 2.1, which is referred to as the Daniel cell. It consists of a zinc (Zn) electrode kept in an aqueous zinc sulfate solution and a copper (Cu) electrode kept in an aqueous copper sulfate ($CuSO_4$) solution.[8] When the switch is closed, zinc is oxidized to zinc ion, giving off two electrons into the solution.

$$Zn \rightarrow Zn^{2+} + 2e^- \tag{2.1}$$

The electrons thus liberated pass through the external circuit, reach the Cu electrode, and reduce a copper ion to copper metal on the surface of the copper electrode.

$$Cu^{2+} + 2e^- \rightarrow Cu \tag{2.2}$$

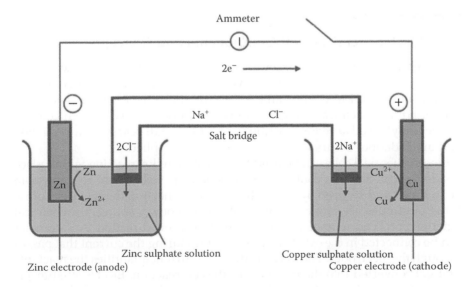

FIGURE 2.1
Schematic representation of a Daniel cell.

The electron flow in the external circuit constitutes an electric current, which is a result of the chemical reaction taking place inside the cell. Ions flow within the electrolytes and across the salt bridge, as shown in the figure, to maintain the ionic charge balance in the solutions. The overall result is the reduction of copper ion by zinc.

$$Cu^{2+} + Zn \rightarrow Cu + Zn^{2+} \tag{2.3}$$

The electrode where oxidation takes place is called the anode and the electrode where reduction takes place is called the cathode. Reactions 1 and 2 are known as half-reactions or redox reactions.

A typical Daniel cell is represented symbolically as

$$Zn \: / \: Zn^{2+}{}_{(aq)}(1M) \: \| \: Cu^{2+}{}_{(aq)}(1M) \: / \: Cu \qquad \qquad A$$

The double vertical line represents a salt bridge. The function of a salt bridge is to bring about contact between two solutions. It is a U-shaped tube bent at right angles and the tube is filled with potassium chloride (KCl), starch, and water and the ends are sealed. The purpose of the salt bridge is to prevent the mixing of electrolyte solution and polarization and to provide a passage for the movement of ions. The concentration of the solution is shown in brackets and the direction of current is represented by an arrow. The left-hand side electrode is the anode and on right-hand side is the cathode.

2.3 Electrochemical Cell Components and Concepts

Electrochemical cells are generally classified as galvanic cells and electrolytic cells. In galvanic cells (also called voltaic cells), chemical energy is converted into electrical energy and in electrolytic cells, electrical energy is converted into chemical energy. Common batteries consist of such cells. In an electrolytic cell, a redox reaction (sum of two half-reactions: a reduction and an oxidation) occurs when electric energy is applied.[9–13]

Electrochemical cells can be used to measure a redox reaction taking place inside the cell and can control these reactions as well. These reactions can be initiated and stopped by connecting or disconnecting the two electrodes. If the electrodes are connected by a battery or a power source, the chemical reactions proceed in its nonspontaneous or reverse direction. An ammeter can be connected in the external circuit for measuring the current that passes through the electrodes. The magnitude of the current quantifies the reactants that get converted into the products in the cell reaction. Electric charge q is measured in coulombs. Faraday, denoted by F, *is* the amount of charge carried by one mole of electrons. For most purposes 1 F equals 96,500 coulombs.

2.4   Thermodynamics of the Electrochemical Cell

By the first law of thermodynamics,

$$\Delta U = q + w \tag{2.4}$$

where ΔU is the change in the internal energy of the system, q is the heat absorbed by the system, and w is the work done by the system. In electrochemistry, the work done is the electrical work, when an electrical charge is moved through an electric potential difference.

Consider a system that undergoes a reversible process at constant temperature and pressure. Here mechanical work done is PV, where P is the pressure and V is the volume and the electrical work done is $w = -P\Delta V + W_{\text{elec}}$. A reversible process at constant temperature and pressure requires q to be equivalent to $T\Delta S$, (where ΔS represents the change in entropy of the system).

Therefore Equation 2.4 becomes,

$$\Delta U_{T,P} = T\Delta S - P\Delta V + W_{\text{elec}} \tag{2.5}$$

At constant pressure, the system's enthalpy change is

$$\Delta Hp = \Delta Up + P\Delta V \tag{2.6}$$

At constant temperature, the Gibbs free energy change is

$$\Delta G_T = \Delta H_T - T\Delta S \tag{2.7}$$

Combining Equations 2.4 and 2.7, we get

$$\Delta G_{T,P} = W_{\text{elec}} \tag{2.8}$$

To obtain a relationship between the electric work and the cell potential, consider an electrochemical cell having two terminals with a potential difference E. The two terminals are connected to the surroundings. When a charge, Q, is moved through a potential difference, E, the work done on the surroundings is EQ. If the charge carriers are electrons, then the total charge,

$$Q = Ne \tag{2.9}$$

where N is the number of electrons and e is the charge. If n is the number of moles of electrons and F is the charge per mole or Faraday constant,

$$Q = nF \tag{2.10}$$

The work done by the system is nFE. If the system transfers energy to the surroundings, the electrical work is negative.

$$\Delta G_{T,P} = W_{elec}$$

Hence,

$$\Delta G_{T,P} = -nFE \tag{2.11}$$

If E is measured in volts, F in coulombs, and n is the number of moles of electrons per mole of reaction, then ΔG will have units of Joules per mole.

2.5 Components of an Electrochemical Cell

2.5.1 Reference Electrodes

Reference electrodes (or standard reference electrodes) are used to determine the potential of other electrodes. The electrode potentials of these electrodes are known. In most electrochemical experiments, the interest lies in only one of the electrode reactions. For this purpose, a reference electrode is used as the other half of the cell. The reference electrode is characterized by its ease to prepare and to be maintained. But the major condition for using an electrode as a reference electrode is that it should have a constant potential. The concentration of any ionic species taking part in the reaction must be held at

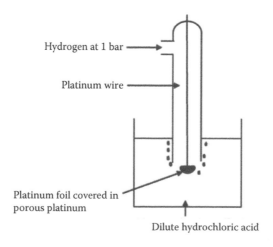

FIGURE 2.2
A standard hydrogen electrode.

a constant value, which can be attained by employing an electrode reaction where a saturated solution of an insoluble salt of the ion is involved.[14-19]

Reference electrodes are generally classified into primary and secondary reference electrodes. The simple example for a primary reference electrode is the hydrogen electrode (Figure 2.2).

By maintaining a constant pressure of hydrogen gas, the potential of a hydrogen electrode can be used for determining the activity of the hydrogen ions in the tested solution. But keeping this activity at unit value is difficult. Any small variation of temperature or a trace amount of soluble impurities alters the pH and the system deviates from the ideal condition.

To solve this issue secondary reference electrodes, are used which are not hydrogen electrodes but they maintain a constant electromotive force (EMF) during the experiment. In practice, potentials are measured against reference electrodes that are easier to work with than the normal hydrogen electrode. Such electrodes are known as secondary reference electrodes; the most common ones are the saturated calomel electrode and silver–silver chloride (Ag–AgCl) electrode.[20,21]

2.5.2 Calomel Electrode

A calomel electrode (represented in Figure 2.3) can be set up by placing small amounts of mercury in a glass vessel, on top of which a thin layer of Hg_2Cl_2 (mercury chloride, also known as calomel) is spread. Calomel is soluble only in water. The two layers of mercury and the calomel are covered with a saturated solution of KCl. A glass tube carrying a platinum electrode is introduced to provide electrical connections. A salt bridge dipped into the KCl solution couples the calomel electrode with other electrodes. The electrode can be used as an anode or cathode to determine the electrode potential of other electrodes.[22] The electrode is symbolically represented as,

$$Hg/Hg^+/KCl \text{ (sat.)} \hspace{4cm} B$$

When the electrode is used as anode,

$$2Hg \rightarrow Hg_2^+ + 2e^- \hspace{4cm} (2.12)$$

$$Hg_2^+ + 2Cl^- \rightarrow Hg_2Cl_2 + 2e^- \hspace{3cm} (2.13)$$

The net reaction is

$$2Hg + 2Cl^- \rightarrow Hg_2Cl_2 + 4e^- \hspace{3cm} (2.14)$$

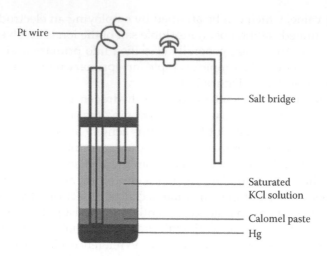

Pt wire

Salt bridge

Saturated
KCl solution

Calomel paste

Hg

FIGURE 2.3
Calomel electrode.

When the electrode is used as cathode,

$$Hg_2Cl_2 + 4e^- \rightarrow 2Hg + 2Cl^- \tag{2.15}$$

The potential of the calomel electrode depends on the concentration of KCl solution in it.

Advantages of the calomel electrode:

1. Ease of construction.
2. Cell potential can be reproduced.
3. Potential remains constant for a longer time.
4. Temperature invariant.

Ag–AgCl electrodes are another example of secondary electrodes. The cell can be represented symbolically as follows:

$$Ag \,|\, AgCl(s) \,|\, Cl^-(aq) \,\|\, Ag(s) + Cl^-(aq)\, AgCl(s) + e^- \qquad\qquad C$$

This electrode is usually a piece of silver wire coated with AgCl. The silver coating is created using silver as the anode in an electrolytic cell containing HCl. The Ag^+ ions combine with Cl^- ions as soon as they are formed at the silver surface.

The potentials of both these electrodes have been very accurately determined against the hydrogen electrode.

2.5.3 pH-Based Reference Electrodes

pH-sensitive electrodes can be used as reference electrodes with a buffer solution of constant pH. The glass electrode is rarely used as the reference electrode, because it is difficult to maintain a constant value for potential and requires frequent standardization of the electrode. Instead, a quinhydrone electrode can be used, whose principle is based on the electrochemically reversible redox system of p-benzoquinone (quinone) and hydroquinone in which hydrogen ions participate. This electrode is simple to construct. Here a noble metal wire, usually platinum, is introduced into the solution containing some crystals of quinhydrone.[23-27]

2.6 Electrolytes

An electrolyte is a substance that splits into ions when dissolved in appropriate ionizing solvents such as water. Soluble salts, acids, and bases are all examples of electrolytes. Electrolyte solutions can also be formed by dissolving some biological and synthetic polymers, characterized by the presence of certain functional groups, which carry a charge.

When a salt is dissolved in a solvent, like water, the individual components of the salt undergoes chemical transformation with the solvent molecules (here water) and will dissociate into ions. The reaction can be explained using the principles of thermodynamics. This process is termed as solvation. Solvation results in the formation of electrolyte solutions. For example, when NaCl is dissolved in water, the salt dissociates into its component ions. The reaction can be expressed as follows:

$$NaCl_{(s)} \rightarrow Na^+_{(aq)} + Cl^-_{(aq)} \tag{2.16}$$

Another possibility for forming electrolytes is when a substance reacts with a solvent and forms ions, as opposed to getting dissociated into its component ions. For example, when carbon dioxide gas reacts with water, it forms a solution containing hydronium carbonate and hydrogen carbonate ions. An electrolyte can exist in a molten state too. A typical example is molten NaCl, which can conduct electricity. An electrolyte in a solution may be described as concentrated if it has a high concentration of ions, or dilute if it has a low concentration of ions.[28-31]

Aqueous electrolytes were employed for electrochemical cells until recently as they provided good ionic conductivity. But they present difficulties in packaging and also report the problem of electrolyte leakage.[32] An alternative to liquid electrolytes is solid polymer electrolytes. In addition to solving the packaging issues, polymer electrolytes have the advantage of being able to be formed into thin films, which enhances the energy density of the cell, which implies they can store more energy. Polyvinyl alcohol is one such type of polymer electrolyte system, which has electrochemical capacitor applications. Another widely used polymer electrolyte is a proton exchange membrane or polymer electrolyte membrane (PEM). A polymer electrolyte system is shown in Figure 2.4. It is basically a semipermeable membrane generally made from ionomers. They conduct protons while being impervious to gases such as oxygen or hydrogen; that is, they separate the reactants while facilitating the transport of protons. This property enables them to be used in a membrane electrode assembly of a proton exchange membrane fuel cell or of a proton exchange membrane electrolyzer. The most widely used and commercially available PEM materials are Nafion fluoropolymers, a DuPont product.[33–35] They make use of a solid polymer membrane (a thin plastic film) as the electrolyte. This polymer is permeable to protons when it is saturated with water, but it does not conduct electrons.

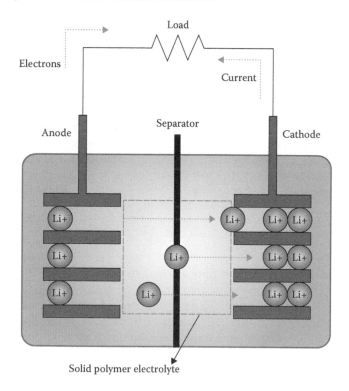

FIGURE 2.4
Supercapacitor based on polymer electrolyte system.

2.7 Electrode Potentials

Current and potential are the two electrical variables of interest in electrochemical cells. Current is related to the rate of the electrode reactions and the potential to the cell energy. A schematic representation of an electrolytic cell is represented in Figure 2.5.

Current is measured in amperes (A), or the amount of electricity in coulombs (C) that passes across a medium per second(s). Potential between the two electrodes is measured in volts (V) with a voltmeter. Potential (V) is the unit of energy or work (J) per amount of electric charge (C). That is, $1\ V = 1\ J/C$ so that the cell potential is a measure of the energy of the cell reaction. The cell is said to be at open circuit when no current flows; that is, when there are no external connections to the electrodes. Under these conditions, no electrode reactions take place.[36]

Measurements of the potentials of galvanic cells at open circuit give information about the thermodynamics of cells and cell reactions. The potential of the cell when the solution concentrations are 1 molar (1M) at

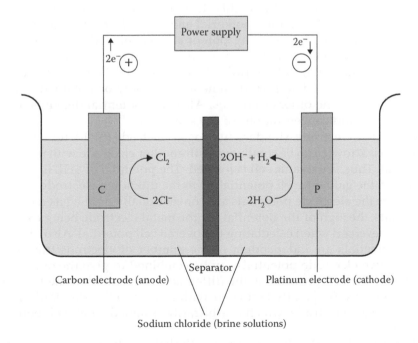

FIGURE 2.5
Schematic representation of an electrolytic cell.

25°C is called the standard potential of the cell and is represented by E°. The available energy (the Gibbs free energy $\Delta G°$) of the cell reaction is related to E° by

$$\Delta G° = -nFE° \qquad (2.17)$$

where n is the number of electrons transferred in the reaction and F is a proportionality constant, called the Faraday (96,485 coulombs/equivalent). The cell potential is the difference in the potential of the two half-cells.

2.8   Calculation of Standard Electrode Potential

In an electrochemical cell, an electrode potential is created between two metals and this potential is available from the redox reactions in the cell. The total cell potential E_{cell} has contributions from both the anode and cathode. The anode has the ability to lose electrons and the potential developed is called oxidation potential and the cathode has the ability to gain electrons and the resulting potential is called reduction potential. Thus, E_{cell} = (reduction potential − oxidation potential). Thus, for the Zn/Cu cell, the voltage of the full cell is: (+0.34 − (−0.76)) = 1.1 V. But this potential cannot be determined alone and can only be obtained in combination with some other electrodes. Also, this potential depends on the temperature and concentrations of the substances.[37–39]

A reference electrode, standard hydrogen electrode (SHE), for which the potential is known, can be combined with another electrode with unknown potential. Thus, a galvanic cell is formed. The potential of SHE is set to be 0.0 V and the galvanic cell potential gives the unknown electrode's potential. Since the electrode potentials are conventionally defined as reduction potentials, the sign of the potential for the metal electrode being oxidized must be reversed when calculating the overall cell potential. Also, the electrode potentials are independent of the number of electrons transferred and the two electrode potentials can be combined to give the overall cell potential. The cell potential is the difference in potential of the two half-cells. These are frequently tabulated with respect to the standard or normal hydrogen electrode, which is arbitrarily assigned a half-cell potential of zero.

Thus, the value, +0.34 V, is assigned to the half-reaction

$$Cu^{2+} + 2e^- \rightarrow Cu \qquad (2.18)$$

Similarly, the standard potential for the Zn/Zn^{2+} cell yields $Zn^{2+} + 2e^- \rightarrow Zn$, which is −0.76V.

Electrochemical cell redox reactions are taking place due to the flow of electrons from anode to cathode. The force on the electrons causing them to move is referred to as electromotive force or EMF (E). One can measure the magnitude of the EMF causing electron flow by measuring the voltage. Electricity is generated due to the electric potential difference between the two electrodes and this difference in potential is called the cell potential. This electric potential varies with temperature, pressure, and concentration of the cell. As mentioned earlier, these redox reactions can be broken down into two half-reactions.[40–43]

2.9 Standard Half-Cell Potential

When a net reaction proceeds in an electrochemical cell, oxidation occurs at one electrode (the anode) and reduction takes place at the other electrode (the cathode). The electrochemical cell consists of two *half-cells* joined together by an external circuit through which electrons flow and an internal pathway that allows ions to migrate between them so as to preserve electroneutrality. Cell potential is the difference between anode and cathode potential:[44–47]

$$E_{cell} = E_{cathode} - E_{anode} \qquad (2.19)$$

when half-reactions are written as reductions.
  Example,

$$Zn \rightarrow Zn^{2+} + 2e^-, \ E_{\frac{Zn}{Zn^{2+}}} = +0.76V \qquad (2.20)$$

$$2H^+ + 2e^- \rightarrow H_2, \ E^0_{(SHE)} = +0.00V \qquad (2.21)$$

The net reaction is,

$$Zn + 2H^+ \rightarrow Zn^{2+} + H_2 \qquad (2.22)$$

$$E_{cell} = -E_{\left(\frac{Zn}{Zn^{2+}}\right)} + E^0_{(SHE)} = -0.76 + 0.00 = -0.76V \qquad (2.23)$$

2.10   Concentration Effects on Cell Potential (Nernst Equation)

Nernst equation relates free energy change in a chemical reaction to the EMF of a cell. To derive this equation, a reduction reaction is considered as a suitable example.

For example, reducing the concentration of Zn^{2+} in the Zn/Cu cell from its standard effective value of $1M$ to a much smaller value:

$$Zn_{(s)}|\ Zn^{2+}_{(aq,.001M)}||\ Cu^{2+}_{(aq)}|\ Cu_{(s)} \qquad\qquad D$$

This will reduce the value of reaction quotient Q for the cell reaction,

$$Zn_{(s)} + Cu^{2+} \rightarrow Zn^{2+} + Cu_{(s)} \qquad\qquad E$$

thus making it more spontaneous and driving it to the right according to Le Châtelier principle, and making its free energy change ΔG more negative than $\Delta G°$, so that E would be more positive than $E°$.[48,49]

According to thermodynamics, equilibrium constant is related to free energy change by van't Hoff's equation as

$$\Delta G = \Delta G^0 + RT \ln Q \qquad\qquad (2.24)$$

which gives

$$-nFE = -nFE^0 + RT \ln Q \qquad\qquad (2.25)$$

which can be rearranged by dividing the whole equation by $-nF$:

$$E = E° - \frac{RT}{nF} \ln Q \qquad\qquad (2.26)$$

Q can be taken as $\dfrac{Zn^{2+}}{Cu^{2+}}$

$$E = E° - \frac{RT}{nF} \ln \frac{Zn^{2+}}{Cu^{2+}} \quad \text{or}$$

$$E = E° + \frac{RT}{nF} \ln \frac{Cu^{2+}}{Zn^{2+}} \qquad\qquad (2.27)$$

By raising the base to ten, the equation will be,

$$E = E° + 2.303 \frac{RT}{nF} \log \frac{Cu^{2+}}{Zn^{2+}} \qquad\qquad (2.28)$$

Substituting the values for R and F at room temperature, the equation changes to,

$$E_n = E° + 0.0591\log\frac{Cu^{2+}}{Zn^{2+}}$$
(2.29)

This is the Nernst *equation* which relates the cell potential to the standard potential and to the activities of the electroactive species. The cell potential will be the same as $E°$ only if $[Cu^{2+}]/[Zn^{2+}]$ is unity.[50]

2.11   Junction Potentials

It is the potential that develops at an interface, or a junction, where a separation of charges is observed. Junction potentials are formed when a metal electrode comes in contact with a solution containing its cation or solutions of varying concentrations, or those containing different substances, when in contact across a junction. A potential can also develop when electrolyte solutions having differing compositions are separated by a boundary, such as a membrane or a salt bridge. The two solutions may contain the same ions, just at different concentrations or may contain different ions altogether. The basic principle is that the ions have different mobilities, which means that they move at different rates. The potential that develops at the junction when two solutions of dissimilar concentrations are brought to contact is termed as liquid junction potential. As is the case with most of the physical or chemical phenomena, the ions in the solution of higher concentration diffuse into the solution having a lower concentration. The rate of this diffusion of each ion will be proportional to its speed when influenced by an electric field (mobility).[51] The speed with which the anion and cation diffusion take place will be different. If the anion diffuses faster than the cation, it will go promptly into the dilute solution. As a result, the dilute solution becomes negatively charged and the concentrated solution becomes positively charged. At the junction of the two solutions, an electrical double layer of positive and negative ions will be produced, creating a potential difference. This potential is called junction potential or diffusion potential, and its magnitude depends on the speed of cations and anions.

2.12   Calculation

The liquid junction potential cannot be measured directly but can be theoretically calculated. The value cannot be accurate because it is impossible to determine experimentally the individual activities of ions of the electrolytes comprising the junction. The EMF of a concentration cell with transference includes the liquid junction potential.[52,53]

$$E_{\text{without transference}} = \frac{RT}{F} \ln \frac{a_2}{a_1} \tag{2.30}$$

where a_1 and a_2 are activities (a parameter which expresses the availability of the ionic species to determine properties, to take part in a chemical reaction or to influence the position of an equilibrium) of HCl in the two solutions, R is the universal gas constant, T is the temperature, and F is Faraday's constant.

$$E_{\text{with transference}} = t_M \frac{RT}{F} \ln \frac{a_2}{a_1} \tag{2.31}$$

where a_2 and a_1 are activities of HCl solutions of right- and left-hand electrodes, respectively, and t_M is the transport number of Cl⁻ (transport number is the fraction of the total current carried by a given ion in an electrolyte).

$$E_{\text{liquid junction potential}} = E_{\text{with transference}} - E_{\text{without transference}}$$

$$= (t_M - 1) \frac{RT}{F} \ln \left(\frac{a_2}{a_1} \right) \tag{2.32}$$

2.13 Elimination

The liquid junction potential affects the exact measurement of EMF of a chemical cell. So it should be eliminated or at least its effect should be minimized. Using a concentrated electrolyte solution in the form of a salt bridge (Figure 2.6), connecting the two cells can minimize the liquid junction potential. A salt bridge is employed because the ions of the salt will be present in excess at the junction and they can carry the current. Usually, a salt bridge consisting of a saturated solution of KCl and ammonium nitrate (NH_4NO_3) with lithium acetate (CH_3COOLi) is placed between the two solutions constituting the junction.[54–58] When such a bridge is used, the ions in the bridge are present in large excess at the junction and they carry almost the whole of the current across the boundary. The efficiency of KCl/NH_4NO_3 is due to the fact that in these salts, the transport numbers of anion and cation are the same and these two ions have approximately equal conductivity.

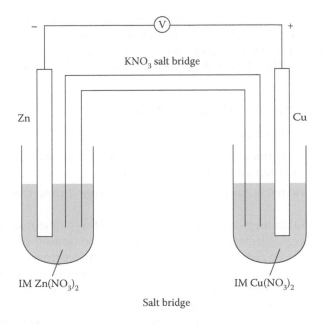

FIGURE 2.6
Salt bridge connecting the two half-cells.

2.14 Mixed Potentials

The value of the potential of a given electrode with respect to a suitable reference electrode, usually SHE, is not always governed by the electrode potential developed by the metal and its ions in solution. The principle of charge conservation requires that there must be at least one reduction and one oxidation in an electrochemical reaction.[59] For example,

$$2H^+ + 2e^- \rightarrow H_2 \qquad (2.33)$$

$$Zn \rightarrow Zn^{2+} + 2e^- \qquad (2.34)$$

Other reactions may also occur when the metal is thermodynamically unstable in aqueous solution. The reduction of hydrogen ions may interfere with the electrode equilibrium. In this example, the metal zinc with negative electrode potential is in an acidic solution containing its cations. If the potential of the zinc electrode is increased, the current density would vary considerably. The standard potential of hydrogen is more positive than that of zinc; the rate of hydrogen discharge increases according to its characteristic Tafel line. At the point where the two Tafel lines meet, both the reactions occur at the same current density.[60–63] The potential at this point is a steady one and is called mixed or corrosion potential. (The relationship between the current flowing and potential for a reaction can be measured, and a straight line relationship should be found. The plots of these straight lines are called Tafel lines.)

2.14.1   Mixed Potential Theory and Corrosion

The mixed potential theory comprises the following facts:[64,65]

1. An electrochemical reaction can be considered to have two or more partial oxidation and reduction reactions.
2. There is no net accumulation of electric charge during the reaction.

That is, a metal immersed in an electrolyte cannot spontaneously accumulate an electric charge. Electrochemical reactions are composed of two or more partial oxidation or reduction reactions. During the corrosion of an electrically isolated metal sample, the total rate of oxidation must equal the total rate of reduction.[66] Corrosion is the tendency of the metal to get destroyed in the surrounding environment through electrochemical reaction. The rate of corrosion depends on the nature of the corrosion product (the oxide layer), which is formed during corrosion. For example, metals such as iron and zinc form oxides during corrosion that are porous and conducting in nature. Hence, the metal will be thoroughly destroyed if the corrosion proceeds.[67–70] On the other hand, metals such as aluminum and titanium form oxides that are nonporous and nonconducting. So corrosion stops as soon as the oxide film is formed. Electrode potential is another rate-determining factor. Whenever two metals are in contact with each other, there exists a potential termed open circuit potential. The rate of corrosion depends upon this potential and higher the open circuit potential, faster will be the corrosion rate.

2.15   Concentration Overpotential

During electrolysis, a minimum potential difference must be applied between the electrodes before decomposition occurs and current flows. For a continuous process, the applied voltage should be higher than the decomposition potential. When decomposition potential is applied, electrolysis may not start. When the potential is increased slightly above the theoretical value, electrolysis will start and this potential is called overpotential (η). In other words, the potential difference brought about by the differences in concentration of the charge carriers between bulk solution and on the electrode surface is called concentration overpotential. It is basically the difference in equilibrium potentials across the diffusion layer and occurs when electrochemical reaction is sufficiently rapid to lower the surface concentration of the charge carriers below that of a bulk solution. It can be stated that concentration overpotential can be taken as the potential of a reference electrode with interfacial concentrations relative to the potential of a similar reference electrode with the concentrations of the bulk solution. For this kind of potential difference measurement, the ohmic potential drop between the electrodes also needs to be considered (ohmic potential drop is the potential due to solution resistance and is the potential required by the ions to move in the solution). The rate of the reaction depends on the ability of the charge carriers to reach the electrode surface and involves the depletion of these carriers at the electrode surface.[9,71,72]

2.16   Electrode–Electrolyte Interface

The electrons are charge carriers present in an electrode, whereas an electrolyte has two types of charge carriers (cations and anions).[73,74] The direction of the current flow is shown in Figure 2.7.

Charge transfer at an interface takes place only due to chemical reactions, as there are no free electrons in the electrolyte and no cations and anions in the electrode as well. If strips of two different metals are put into an appropriate electrolyte solution, one is likely to find that a potential difference appears between the two strips. This system, as shown earlier, is a galvanic cell commonly known as a battery. The associated EMF of the chemical reaction

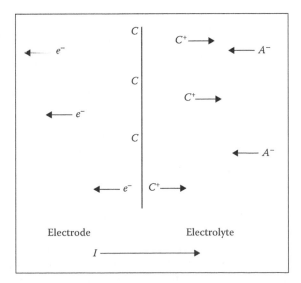

FIGURE 2.7
Charge transport at the interface between the electrode and the electrolyte.

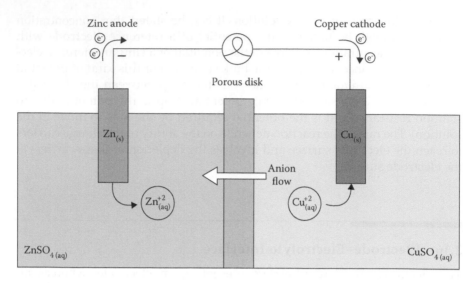

FIGURE 2.8
Schematic representation of a galvanic cell.

produces a potential difference between both the metals and the electrolyte. The interface between each piece of the metal and electrolyte solution forms an electrode. Figure 2.8 gives the schematic representation of a galvanic cell.

Electrodes are always used in pairs. It is not possible to make voltage measurements or pass current into an electrolyte without using two electrodes.

A simple example is a piece of Cu and a piece of Zn both partly immersed in the same dilute solution of $CuSO_4$. The copper becomes about 1 volt positive with respect to the zinc. The potential change occurs in two steps at the electrode–electrolyte interfaces, not in the bulk solution.

Both metals are negative with respect to the electrolyte, but the potential of the zinc is more negative when compared with that of the copper. It is impossible to measure these absolute electrode potentials directly because contact cannot be made with the electrolyte without using another electrode. However the cell potential, which is the difference between the two electrode potentials, can be measured directly.

In all electrochemical experiments, a metal–electrolyte interface is always present. The metal in contact with the electrolyte is called the electrode. There are different types of charge transport phenomena taking place in the electrolyte or in the electrode as represented in Figure 2.9. But the movement of charged species on the metal electrolyte interface is an entirely different behavior. The charge carriers from the solute are not able to migrate into the metal and the electrons from the metal cannot migrate into the solute without assistance. Electric current can pass from the metal into the solute through a charge exchange at the interface.[18] This happens through an electron transfer from the electrode to the ions

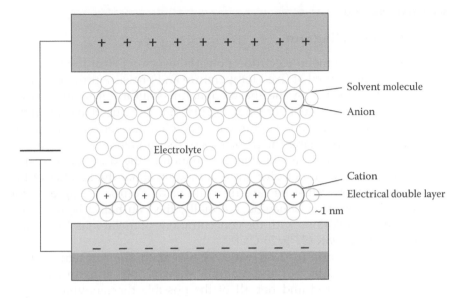

FIGURE 2.9
Transport at the electrode–electrolyte interface and mobile charged species. (From Gregory, D.P., *Modern Aspects of Electrochemistry*, 10, 239, 1975.)

in the liquid. This exchange can be utilized with a redox reaction at the surface of the metal electrode. Every time an electrode is brought into contact with an electrolyte, these redox reactions take place. This results in the accumulation of certain net charge in the electrode–electrolyte interface and builds up a charge separation on the interface, thus preventing further redox reactions.

There are many types of electrodes:

- *Inert metal electrodes or capacitors:* These metals exhibit high resistance to redox reactions and do not allow charge transfer between the two phases.
- *Nonpolarizable electrodes:* These electrodes are characterized by the absence of charge accumulation on their surface. This is owing to the fact that they allow charge transfer from the electrode to the solute.
- *Partially polarizable electrodes:* These electrodes exhibit varying levels of charge barriers to the movement of charged particles. Usually, metal–semiconductor electrodes behave as polarizable or non-polarizable over a range of potentials.

Equilibrium potential describes the electric transport through the electrode interface. An electrode kept in contact with an electrolyte will facilitate redox reactions and will lead to the formation of a net charge separation at the interface between the electrode and the electrolyte. As a result of this charge separation, an opposing electric field is built up at the interface, which prevents further redox reactions. This potential can be measured by using a standard hydrogen electrode.

2.17 Pourbaix Diagram

Graphs of reversible metal electrode potentials versus pH of the solutions in which they are dipped at fixed temperature and pressure provide important information regarding the thermodynamic stability of many phases. Pourbaix diagram is a plot of potential and pH and can be used to explain the corrosion reactions of a metal kept in aqueous solutions. Pourbaix diagrams can be constructed by using the principles of thermodynamics and Nernst equation. They can provide valuable information in the study of corrosion phenomena. But they cannot give any information about the kinetics of the reactions. Hence, one must be careful while using Pourbaix diagrams for explaining the corrosion behavior. Normally, the Pourbaix diagrams are built for aqueous solutions with the concentrations of metal ions $10^{-6}M$ and at the temperature 298K (77°F/25°C).[75]

The Pourbaix diagram for iron–water system at an ambient temperature is shown in Figure 2.10. For the diagram shown, only anhydrous oxide species were considered and not all of the possible thermodynamic species are shown. Here the dashed lines, (a&b) give the pH dependence of the equilibrium potentials of hydrogen and oxygen electrodes. The area in the graph indicated as "passivation" corresponds to the formation of solid compound (iron oxide) on the metal surface. This oxide layer protects the metal from corrosion. Iron oxidizes in this zone but the resulting iron oxide film prevents further oxidation, thereby protecting the metal underneath. This

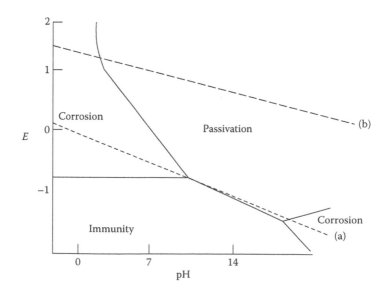

FIGURE 2.10
Pourbaix diagram of iron in aqueous solution.

is termed as passivation. The area marked as "immunity" represents the electrochemical reaction of reduction of the iron ions. No corrosion occurs in this zone. Metallic iron oxidizes in the area marked as "corrosion" zone. Horizontal lines of the Pourbaix diagrams represent the redox reactions, which are independent of the pH. Vertical lines of the diagrams represent the non–redox reactions. Such non–redox reactions do not involve electrons and are dependent on pH. The diagonal lines correspond to the redox reactions, which are dependent on the pH (of the electrolyte).

Limitations of Pourbaix diagrams:

- Corrosion kinetics cannot be studied using these diagrams.
- The diagrams are derived for only specific temperature and pressure conditions.
- The diagrams are derived for selected concentrations of ionic species.
- Most diagrams consider pure substances only.
- The diagram does not consider the nonideal behavior of aqueous solutions.

2.18 Activity Coefficients

The interaction between the ions influences the properties they exhibit. The ionic species are surrounded by oppositely charged ions, thereby forming a cloud around it. If the central ion has to react in the solution or at an electrode, it should free itself from this cloud. For this, energy is required, and the overall result is the loss of energy to the system. This is followed by the loss of activity of the ion, which increases with the concentration. A parameter, which expresses the availability of the ionic species to determine properties, to take part in a chemical reaction, or to influence the position of equilibrium, is called activity (a) and is associated with concentration (c) by the relation,

$$a_i = \gamma_i c_i \tag{2.35}$$

where γ_i is known as activity coefficient, which depends on the unit in which concentrations for a given system are expressed, that is, as molarity, molality, or mole fraction. Thus, the chemical potential (μ_i) of a species may be expressed as,

$$\mu_i = \mu_i^\circ + RT \ln \chi_i \gamma_i \tag{2.36}$$

where χ_i is the mole fraction and γ_i is the activity coefficient.

2.19   Electrochemical Kinetics

Electrochemical kinetics is a branch of electrochemistry dealing with the rate of electrochemical processes. The electrochemical reactions taking place at the interface between an electrode and an electrolyte influences the net reaction rate. The current that flows between the anode and the cathode as a result of the chemical reactions is a measure of the rate of the reaction. Electrochemical corrosion refers to the deterioration of a metal by a chemical process. The resulting mass loss of the metal can be determined using Faraday's law, which relates the current flowing to the mass of metal corroded:[76]

$$\text{Mass}_{\text{loss}} = \frac{(I \times t \times M)}{z \times F} \tag{2.37}$$

where I is the current flowing in amperes, t is the time in seconds, M is the molar mass of the material, z is the number of electrons involved in the reaction, and F is Faraday's constant.

2.20   Charge Transfer Mechanisms

The reasons for the thermodynamically irreversible behavior of the reactions at the electrode interface can be observed in the elementary act of the charge transfer. A typical electrode reaction comprises the transfer of charge between an electrode and a species in the solution. This charge transfer act is inhibited by the presence of an energy barrier between the oxidized and the reduced state. This barrier indicates that the redox reaction could occur only in certain conditions. During the course of numerous interactions with other species (atoms, ions, molecules, etc.), a molecule can be excited to a state in which it has an abnormal energy content. This energy content is sufficient for most of the species in a chemical reaction to come into the transition state. The transition state represents the top of the energy barrier, an event before a reaction is triggered. If such a probabilistic model is applied for an event like electron transfer at an interface, the mathematical calculations show that the electron exchange reactions at the electrodes would be impedingly slow. Quantum mechanically speaking, for a fast electron exchange to happen, electrons in a particle outside the double layer should attain certain well-defined quantized energy states equal to those in which free electrons exist in the metal. These energy states can be attained by the particle at a lower energy content than that needed for its transfer over the top of the energy barrier. This process of electron exchange between the electrode and a particle in solution is called electron tunneling.

The rate of chemical processes can be influenced by modulating the concentrations of reactants or by changing the temperature or both; the rate of electrochemical reactions also can be changed by varying the electrode potential. Making the electrode more electronegative increases the number of electrons in the metal that are ready to tunnel to ions, thereby increasing the rate of the reduction process. However, making the electrode more electropositive decreases this rate and enhances the number of particles ready to donate electrons, thus increasing the rate of the oxidation process.[77]

Thus, it is possible to have a correlation between the rate of reaction and the concentration of the reacting species and also between the rate of reaction and the electrode potential where a direct proportionality could be established. At any electrode potential, redox reactions are taking place, but at different rates; the rate of each reaction is governed by the respective concentration and by the corresponding effects of the potential. This rate of an electrochemical reaction is represented as the electric current density, which is the measure of the quantity of electrons moving in a certain volume of space during a specified unit of time. One of the methods to evaluate the charge transfer mechanism is through cyclic voltammetry. This method offers a rapid measurement of the redox potentials of the electroactive species. The potential of a working electrode (electroactive species) is measured against a reference electrode (platinum foil/wire), which maintains a constant potential.

2.21 Butler–Volmer Equation

The Butler–Volmer equation gives the fundamental relationship in electrochemistry describing how the electrical current on an electrode responds to the changes in the potential (considering that redox reactions can occur on the same electrode).[78] The Butler–Volmer equation can be written as:

$$I = Ai_0 \left\{ \exp^{\left(\frac{\alpha_a \eta F}{RT} \right)} - \exp^{\left(-\frac{\alpha_c \eta F}{RT} \right)} \right\} \qquad (2.38)$$

where I is electrode current in amperes (A), i_0 is exchange current density in A/m^2, η is activation overpotential defined as $(\eta = E - E_{eq})$, in which E is the electrode potential and E_{eq} is the potential at the equilibrium, α_a and α_c are the cathodic and anodic charge transfer coefficients respectively; F is the Faraday constant, R is the universal gas constant.

2.22   Batteries and Supercapacitors

One of the oldest and most important applications of electrochemistry is the storage and conversion of energy. To sum up the discussion, a cell is a single arrangement of two electrodes and an electrolytic solution within the cell that produces a chemical reaction when electricity is passed. Each cell is made of two electrodes; one that liberates electrons is the oxidizing electrode and the other that absorbs electrons is the reducing electrode. A galvanic cell converts chemical energy to work and an electrolytic cell converts electrical work into chemical energy.

Devices that carry out these conversions are called batteries. It is a combination of two or more cells arranged parallelly or in series. Primary batteries are commonly referred to as dry cells. In ordinary batteries, the chemical components are contained within the device itself. If the reactants are supplied from an external source as they are consumed, the device is called a fuel cell.[77]

A potential energy is generated by the movement of electrons, which can be used in a circuit.[33] Supercapacitors are the capacitor-based storing cells, which are more efficient than the normal batteries. Indifference to temperature change, shocks, and vibrations and a long service life are some of the other advantages of supercapacitors. A disadvantage is that the storage capacity of the supercapacitor is restricted due to limitations on the size of the electrodes. As a result, supercapacitors are larger than a battery of the same capacity.

The supercapacitor, also known as ultracapacitor or double-layer capacitor, differs from a regular capacitor in that it has a very high capacitance. It is capable of charging and storing energy at a higher density than standard capacitors. A capacitor stores energy by the means of static charges as opposed to an electrochemical reaction as is the in case in batteries. Applying a voltage differential on the positive and negative plates charges the capacitor. The modern supercapacitor uses special electrodes and electrolytes. The electric double-layer capacitor is the most common system in use today. It is a carbon-based system and uses an organic electrolyte that is easy to manufacture.

The higher energy density of supercapacitors over capacitors provides access to new power electronic and industrial storage applications. Many automotive companies use double-layer capacitors to shield certain electrical engine parts from voltage fluctuations. The rapid charging rate of the supercapacitors also makes them effective in mass transit braking mechanisms and portable fuel cells for electric/hybrid vehicles. The newly proposed applications for large size supercapacitors include load leveling in electric and hybrid vehicles as well as in the traction domain, the starting of engines, applications in the telecommunication and power quality, and reliability requirements for uninterruptable power supply or UPS installations. Supercapacitors also serve as backups to primary batteries to bridge brief power interruptions or for a smooth electrical flow. To increase the range of applications of supercapacitors, newer forms of dielectric materials are now being developed. Electrodes based on carbon nanotubes (CNTs) provide very high power and energy performance because of the higher surface area, conductivity, and the ability of CNTs to optimize capacitor properties. Supercapacitors using CNTs as electrode can potentially achieve higher energy densities than conventional capacitors with an equivalent amount of delivered power. So combining a supercapacitor with alternative energy sources can replace car batteries and can help the environment by going green.[76]

2.23 Differences between a Supercapacitor and a Battery

In a battery, energy is stored in the form of chemical energy, whereas in a capacitor, the energy is stored as an electrostatic field. A battery stores a lot of energy and less power, whereas a capacitor can provide large amounts of power, but stores less energy. The most important difference between a supercapacitor and a battery is the principle of electrochemical energy storage. Electrochemical energy can be stored in two fundamentally different ways. In a battery, the potentially available chemical energy storage requires oxidation and reduction of reagents to release charges that can perform electric work when they flow between two electrodes having different potentials. Thus, the charge storage is achieved by electron transfer that produces a redox reaction in the electroactive material according to Faraday's law. The main advantages and limitations of supercapacitors are shown in Table 2.1.

TABLE 2.1

Advantages and Limitations of Supercapacitors

Advantages	Limitations
Unlimited life	High self-discharge
High specific power	Low specific energy
Charges very easily and simple charging methods needed	Low cell voltage
Charge and discharge at low temperatures	High cost

References

1. Bard AJ, Faulkner LR. *Electrochemical Methods*, 2nd edn. India: Wiley India Ltd., 2010.
2. Rieger PH. *Electrochemistry*, 2nd edn. New York, USA: Springer, 1993.
3. Crow DR. *Principles and Applications of Electrochemistry*, 4th edn. Boca Raton, FL: Chapman & Hall/CRC, 1994.
4. Cottrell FG, Z. Residual current in galvanic polarization, regarded as a diffusion problem. *Phys Chem* 1903; 42: 385.
5. O'M. Bockris J, Reddy AKN, Gamboa-Aldeco M. *Mod Electrochem 2A*. Berlin, Germany: Springer, 2006.
6. Vijayasarathy PR. *Engineering Chemistry*. New Delhi, India: PHI Learning Pvt. Ltd., 2011.
7. Sharma YR. *A New Course in Chemistry (Solution Chemistry)*. India: Kalyani Publishers, 2011.
8. Zhiwei Y, Decio C, Fangxia F, John PF, Kenneth J. Balkus Jr. Novel inorganic/organic hybrid electrolyte membrane. *Fuel Chem* 2004; 49(2): 600.
9. O'M. Bockris J, Reddy AKN. *Modern Electrochemistry 1 (Ionics)*. Springer: 2006.
10. Narayan R, Viswanathan B. *Chemical and Electrochemical Energy Systems*. New Delhi, India: Universities Press, 1997.
11. Newman JS, Tobias CW. Theoretical analysis of current distribution in porous electrodes. *J Electrochem Soc* 1962; 109(12): 1187.
12. Sharma BK. *Electrochemistry*, Meerut, India: Goel Publishing House, 1976.
13. Satish S, Sindhwani KL. *Electrochemistry*, Meerut, India: Jai Prakash Nath & Co., 1976.
14. Glasstone S. *An Introduction to Electrochemistry*, New Delhi, India: East-West Press Private Limited, 2005.
15. Hamann CH, Hamnett A, Vielstich W. *Electrochemistry*, Germany: Wiley-VCH, 1998.
16. Khosla BD, Miglani S, Gulati A. *Physical Chemistry*, India: R. Chand & Company, 1997.
17. Barrow GM. *Physical Chemistry*, 5th edn. New Delhi, India: Tata McGraw-Hill Publishing Company Limited, 1992.
18. Moore WJ. *Physical Chemistry*, 5th edn. Hyderabad: Orient Longman, 1993.
19. Puri BR, Sharma LR, Pathania MS. *Principles of Physical Chemistry*, New Delhi, India: Shoban Lal Nagin Chand & Co., 1986.
20. Murray RW. Chemically modified electrodes. *Electroanal Chem* 1984; 13: 191.
21. Gregory DP. *Mod Aspects Electrochem*, 1975; 10: 239.
22. Christensen PA, Hamnett H. *Techniques and Mechanisms in Electrochemistry*, New York: Blackie Academic and Professional, 1994.
23. Koryta J, Dvorak J, Kavan L. *Principles of Electrochemistry*, 2nd edn. New York: Wiley, 1993.
24. Khalid MAA. *Electrochemistry*, Rijeka, India: InTech online edition, 2013.
25. Franceschetti DR, MacDonald JR. Diffusion of neutral and charged species under small-signal A. C. conditions. *J Electroanal Chem* 1979; 101: 307.
26. Broussely M, Biensan P, Simon B. Lithium insertion into host materials: The key to success for Li ion batteries. *Electrochem Acta* 1999; 45: 3.
27. Reiger PH. *Electrochemistry*, 2nd edn. New York: Chapman and Hall, 1994.

28. Newman JS. *Electrochemical Systems*, 2nd edn. New Jersey: Prentice-Hall, 1991.
29. Gauchia L, SanzDynamic J. *Modeling of Electrochemical Energy Systems*, Lambert Academic Publishing, 2010.
30. Chun JH. *Developments in Electrochemistry*, Rijeka, India: InTech Publishers, 2012.
31. Sur UK. *Recent Trend in Electrochemical Science and Technology*, Rijeka, India, 2012.
32. Peng C, Zhang S, Jewell D, et al. *Progress Nat Sci*, China: CAS, 2008; 18: 777.
33. Conway BE, Birss V, Wojtowicz J. The role and utilization of pseudocapacitance for energy storage by supercapacitors. *J Power Sources* 1997; 66: 1–14.
34. Eaves S, Eaves J. A cost of comparison of fuel cell and battery electric vehicle. *J Power Sources* 2004; 130: 208–210.
35. Chen JH, Li WZ, Wang DZ. Carbon nanotube and conducting polymer composites for supercapacitors. *Carbon* 2002; 40: 1193–1197.
36. Li C, Wang D, Liang T. Imaging the two gaps of the high-temperature super-conductor Bi2Sr2CuO6+x. *Mater Lett* 2004; 58: 3774–3778.
37. Lee JY, Liang K, An KH, Nickel oxide/carbon nanotubes nanocomposite for electrochemical capacitance. *Synth Met* 2005; 150: 153–157.
38. Laforgue A, Simon P, Fauvarque JF. High power density electrodes for carbon supercapacitor applications. *J Electrochem Soc* 2003; 150: A645–A649.
39. Gupta V, Miura N. High performance electrochemical super capacitor from electrochemically synthesized nanostructured polyaniline. *Mater Lett* 2006; 60: 1466–1469.
40. Kotz R, Carlen M. Principles and applications of electrochemical capacitors. *Electrochim Acta* 2000; 45: 2483–2487.
41. Conway BE. *Electrochemical Supercapacitor: Scientific Fundamentals and Technological Application*, New York: Kluwer Academic/Plenum Publishers, 1999.
42. Christensen PA, Hamnett A. *Techniques and Mechanisms in Electrochemistry*, London: Chapman and Hall, 1994.
43. De Bruijn F. The current status of fuel cell technology for mobile and stationary applications. *Green Chem* 2005; 7: 132–150.
44. Winter M, Brodd RJ. What are batteries, fuel cells and supercapacitors? *Chem Rev* 2004; 104: 4245–4269.
45. Ohzuku T, Ueda A. *J Electrochem Soc* 1994; 141: 2972.
46. Peramunage D, Abraham KM. Preparation and electrochemical characterization of overlithiated spinel LiMn2O4. *J Electrochem Soc* 1998; 145: 1131–1135.
47. Fong R, Von Sacken U, Dahn JR. Studies of lithium intercalation into carbons using nonaqueous electrochemical cells. *J Electrochem Soc* 1990; 137: 2009.
48. Kwon OJ, Park CK, Kim TY, et al. A study of increasing regeneration energy using electric double layer capacitor. *WHEC Lyon France* 2006; 16: 13–16.
49. Koppel T. *Powering the Future*. Toronto, USA: John Wiley and Sons, 1999.
50. Ranga S. Jayashree, Gancs L, Eric R, et al. Air-breathing laminar flow-based microfluidic fuel cell. *J Am Chem Soc* 2005; 127: 16758–16759.
51. Ding Y, Chen M, Erlebacher J. Metallic mesoporous nano composites for electro catalysis. *J Am Chem Soc* 2004; 126: 6876–6877.
52. Lux KW, Rodriquez KJ. Template synthesis of arrays of nano fuel cells. *Nano Lett* 2006; 6: 288–295.
53. Atkins PW, Jones P, Loretta. *Chemical Principles*, 3rd edn. California, USA: W.H. Freeman and Company, 2005.

54. Padma Kumar P, Yashonath S. Ionic conduction in the solid state. *J Chem Sci* 2006; 118(1): 135–154.
55. Padma Kumar P, Yashonath S. Ion mobility and levitation effect: anomalous diffusion in nasicon-type structure. *J Phys Chem* 2002; 3443: B106.
56. Alamo J. *Solid State Ionics* 1993; 547: 63–65.
57. Sobha KC Rao KJ. Luminescence of, and energy transfer between Dy3+ and Tb3+ in NASICON-type phosphate glasses. *J Phys Chem Solids* 1996; 57: 1263.
58. Sobha KC, Rao KJ. 31P MAS NMR investigations of crystalline and glassy NASICON-type phosphates. *J Solid State Chem* 1996; 121: 197.
59. Maier J. Nano ionics: ion transport and electrochemical storage in confined systems. *Nature Mater* 2005; 11: 805–815.
60. Despotuli AL, Andreeva AV. *Mod Electron* 2007; 7: 24–29.
61. Despotuli,AL, Andreeva AV, Rambabu B. *Ionics* 2005; 11(3–4): 306–314.
62. Yamaguchi S. *Sci Techn Adv Mater* 2007; 8(6): 503.
63. Jinshan Y, Yi D, Caixia X, et al. Nanoporous metals by dealloying multi-component metallic glasses. *Chem Mater* 2008; 20: 4548–4550
64. Park KW, Sung YE, Han S, et al. Origin of the enhanced catalytic activity of carbon nanocoil-supported PtRu alloy electro catalysts. *J Phys Chem* 2004; 108(204): 939–944.
65. Boyea JM, Camacho RE, Turano SP, et al. Carbon nanotube-based supercapacitors: Technologies and markets. *Nanotechnology Law & Business* 2007; 585–593.
66. Srinivasan S. *Fuel Cells—From Fundamentals to Applications*. New York, USA: Springer, 2006.
67. Bard A, Faulkner L. *Electrochemical Methods: Fundamentals and Applications*, 2nd edn. John Wiley and Sons, Inc., 2001.
68. Chang BY, Park SM. Integrated description of electrode/electrolyte interfaces based on equivalent circuits and its verification using impedance measurements. *Anal Chem* 2006; 78: 1052–1106.
69. Bockris JM, Reddy AKN. *Modern Electrochemistry*. New York: Plenum, 1998.
70. Oldham KB, Myland JC. *Fundamentals of Electrochemical Science*. New York: Academic Press, 1994.
71. Rieger, Philip H. *Electrochemistry*, 2nd edn. New York: Chapman and Hall, 1994.
72. Atkins PW. *Physical Chemistry*. Walton Street, Oxford, UK: Oxford University Press, 1978.
73. Ives DJG. *Chemical Thermodynamics*. London, UK: University Chemistry, Macdonald Technical and Scientific, 1971.
74. O'Donnell, M., Electrodes. BME/ECE 4/517, U. Mich Webster. http://www.spx.arizona.edu/Classes/BME517/lecture/electrodes_ppt.pdf
75. Yang Z, Coutinho D, Feng F, et al. Novel inorganic/organic hybrid electrolyte membranes. *Prepr Pap Am Chem Soc Div Fuel Chem* 2004; 49: 599.
76. Scholz F. *Electrochemistry—Science for the Future*. Universität Greifswald. Germany: Springer-Verlag Berlin Heidelberg, 2008.
77. Perez N. *Electrochemistry and Corrosion Science*. Mass. USA: Kluwer Academic Publishers, 2004.
78. Brookins DG. *Eh-pH Diagrams for Geochemistry*. New York, SA: Springer-Verlag, 1988.

3

Supercapacitors: Fundamental Aspects

S. Jayalekshmi and Anand Puthirath

CONTENTS

3.1   Introduction

The breathtaking technological innovations in all walks of life continue to make human life on earth more and more comfortable day by day. However, the reality that the existing energy resources fueling these technologies are being exhausted at an alarmingly faster pace should not be overlooked. The biggest challenge before human society in the present century is to develop and maintain sustainable and green energy resources to support comfortable human life for many, many future generations. Practical realization of renewable and efficient energy sources is the urgent need of the hour. When it comes to energy generation, the significance of energy storage follows naturally.

Capacitors are fundamental electrical circuit elements that store electrical energy in the order of fractions or multiples of farads [one farad is the charge in coulombs that a capacitor will acquire when the potential across it changes by one volt (V)]. Capacitors have two main applications. The first is the function by which a material can be charged with electrical energy, which can also be discharged when required. This function is applied to smoothing circuits of power supplies, backup circuits of microcomputers, and timer circuits that make use of the periods of charging or discharging. The other application is to block the flow of direct current. This function is applied in filters that extract or eliminate particular frequencies. This is indispensable to circuits where excellent frequency characteristics are required.

As mentioned earlier, in response to the changing global requirements, energy has become the primary focus of the major world powers and the global scientific community. Consequently, in addition to the extensive research activities for realizing efficient and newer energy sources, much interest has also been focused in developing and refining more efficient energy storage devices. One of such storage devices, the supercapacitor, has matured significantly in its design and functioning, over the last decade and has emerged with the potential to facilitate major advances in energy storage.

Supercapacitors, also referred to as ultracapacitors or electrochemical capacitors, are characterized by electrode materials having a very high surface area and thin electrolytic dielectric materials. As will be seen later, this leads to an increase in both capacitance and energy. They have capacitance values that are several times greater than the values shown by conventional capacitors. Such high values of capacitance enable the supercapacitors to show high energy densities in addition to the high power density exhibited by conventional capacitors. A brief background regarding the working principles of capacitors will be helpful to get a better understanding of the nature of supercapacitors.[1-6]

3.2 Electrostatic Capacitor

Ordinary electrostatic capacitors, shown schematically in Figure 3.1, consist of two conducting electrodes separated by an insulating dielectric material. When a certain amount of voltage is applied across the two electrodes, opposite charges build up on the surface of each electrode. The function of the dielectric material is to keep the charges separated. The separation of opposite charges gives rise to an electrical field that allows the capacitor to store electrical energy. The ability of a capacitor to store electrical energy is represented by its capacitance, C.

C is defined as the ratio of the stored (positive) charge Q to applied voltage V:

$$C = \frac{Q}{V} \tag{3.1}$$

For a conventional capacitor, C is directly proportional to the surface area A of each electrode and inversely proportional to the distance between the electrodes d:

$$C = \varepsilon_0 \varepsilon_r \frac{A}{d} \tag{3.2}$$

where ε_0 is the dielectric constant (or "permittivity") of free space and ε_r, the dielectric constant of the insulating material between the

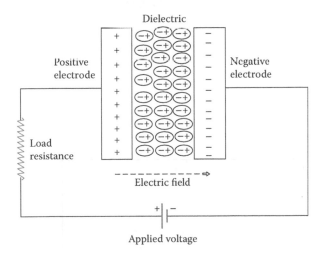

FIGURE 3.1
Schematic representation of electrostatic capacitors.

electrodes. The product of the two parameters constitutes the constant of proportionality. The variation in the voltage V and the electric field E of a capacitor with the separation between the electrode plates d is shown graphically in Figure 3.2. A capacitor is characterized by its energy density and power density. These parameters can be calculated per unit mass or per unit volume.

The expression for the energy E_n stored in a capacitor is given below. As seen from the equation, the stored energy is directly proportional to the capacitance value.

$$E_n = \frac{1}{2}CV^2 \tag{3.3}$$

The power P of a capacitor is the energy utilized or extracted per unit time. P is calculated by considering the capacitor as a circuit element in series with an external "load" resistance R. This resistance also includes that which is offered by the internal elements of a capacitor, such as current collectors, electrodes, and dielectric material. The resistance measured is the aggregate resistance of all these factors and is termed as the equivalent series resistance (ESR). The voltage during discharge is determined by all these resistance factors. When measured at matched impedance ($R = $ ESR), the maximum power P_{max} for a capacitor is given by the equation:

$$P_{max} = \frac{V^2}{4 \times ESR} \tag{3.4}$$

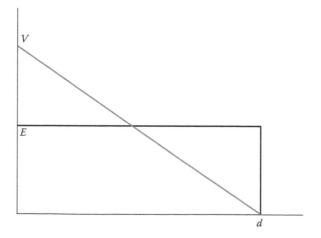

FIGURE 3.2
Graphical representation of the variation of voltage (V) and electric field (E) of a capacitor with the distance of separation between the electrode plates (d).

This relationship explains how the ESR limits the maximum power of a capacitor. The characteristic feature of a capacitor is that it offers a very high power density, but low energy density. On the other hand, electrochemical batteries and fuel cells show very high energy density, even though they have lower power densities. This means that a battery can store more total energy than a capacitor, but it cannot deliver it very quickly, as its power density is low. Capacitors, on the other hand, store relatively less energy per unit mass, but the stored energy can be discharged rapidly to produce a lot of power and hence, their power density is usually high.[4,7]

3.3   Electrolytic Capacitor

The next-generation capacitors are the so-called electrolytic capacitors. These capacitors are assembled using aluminum and tantalum as electrodes and ceramic materials as dielectric media. By architecture, in these capacitors, solid/liquid electrolytes with a separator between two symmetrical electrodes are used, as shown schematically in Figure 3.3. Electrolytic capacitors use an electrolyte as a conductor between the dielectric medium and an electrode. A typical aluminum electrolytic capacitor includes an anode foil and a cathode foil, both made of aluminum, processed by surface enlargement and formation treatments. The anode and cathode foils are made up of high-purity, thin aluminum, having a thickness of 0.02–0.1 mm. As mentioned earlier, increasing the plate area will lead to an increase in the capacitance.

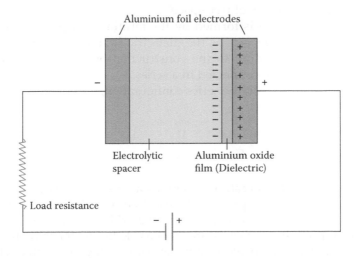

FIGURE 3.3
Schematic representation of an electrolytic capacitor.

This can be achieved by increasing the surface area that is in contact with the electrolyte. Usually this is done by etching the foils. As a result of etching, the aluminum dissolves and a dense network of billions of microscopic tunnels, which penetrate through the foil, is created.

The dielectric is attached to, or contained in, the anode foil. It is a thin layer of aluminum oxide, Al_2O_3, which is chemically grown on the anode foil during a process called "formation." The negative aluminum foil, or cathode, is provided with an electrolyte solution, which serves the following two purposes:

- Creates good contact with the anode by permeating all through the etched structure.
- Repairs any flaws in the oxide dielectric layer when the capacitor is polarized.

The etched structure of the foil ensures good contact with the electrolyte, thereby reducing the series resistance. Anode foil has a thin stabilized oxide film, which serves as the dielectric medium. Such architecture will possess a very high capacitance.

The capacitance of the dielectric portion of the anode aluminum foil can be calculated from Equation 3.2 as follows:

$$C_a = 8.85 \times 10^{-8} \frac{\varepsilon S}{d} \tag{3.5}$$

Here ε is the dielectric constant, S is the surface area (cm^2), and d is the thickness (cm) of the dielectric portion and the numerical constant corresponds to the permittivity of free space.

The oxide film, which is formed by the forming voltage (1 V or less) or naturally during storage, acts as the dielectric for the cathode foil that has a capacitance C_c. According to the construction of aluminum electrolytic capacitors, C_a and C_c are connected in a series.

As these capacitors are in the series configuration, the effective capacitance is given by the equation:

$$\frac{1}{C_{total}} = \frac{1}{C_{anode}} + \frac{1}{C_{cathode}} \tag{3.6}$$

The patterned anode and cathode foils have a considerably large surface area. Due to the excellent contact provided by the electrolyte, the separation between the cathode and the anode constituted by the oxide dielectric layer is significantly small. These prime factors are responsible for the much higher capacitance (of the order of μF) of electrolytic capacitors compared with electrostatic capacitors (of the order of pF). Since electrolytic capacitors exhibit exceptionally high capacitance, they are already commercialized in full scale.

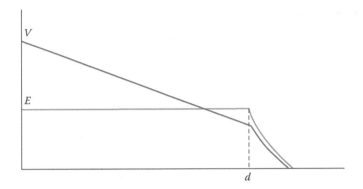

FIGURE 3.4
Graphical representation of variation of voltage (*V*) and electric field (*E*) of an electrolytic capacitor.

The variation in voltage (*V*) and electric field (*E*) of an electrolytic capacitor with the distance of separation between the cathode and aluminum oxide film (dielectric) *d* is graphically represented in Figure 3.4. It can be seen that beyond *d*, electric field *E* and voltage *V* show a curved decrease.[3,4,8]

3.4 Electrical Double-Layer Capacitor

The third-generation capacitors are constituted by an electric double-layer capacitor (EDLC), where the electrical charge stored at a metal–electrolyte interface is the driving force behind the construction of a storage device. The interface can store electrical charge of the order of approximately 10^6 F. The main constituent in the electrode construction is activated carbon. In spite of being known for almost 40 years, the technology was not well researched until recent times. The need for this revival of interest has arisen due to the increasing demands for electrical energy storage in applications, such as digital electronic devices, implantable medical devices, and stop/start operation in vehicle traction, which need very short, high-power pulses that can be realized by EDLCs. Similar to batteries, they possess high energy density in addition to quite high power density. They also possess a longer life cycle than batteries and a higher energy density than the age-old ordinary capacitors. The combination of all these features has led to the evolution of a new breed of storage devices called hybrid charge storage devices, which are commercialized and used widely, in which an electrochemical capacitor is interfaced with a fuel cell or a battery. In these capacitors, both anode and cathode are made up of carbon, along with organic and aqueous electrolytes.[3,9,10] The new device, *supercapacitor*, is a common variation of the term EDLC.

3.5   Technological Aspects of Supercapacitors

3.5.1   Construction

The basic components of a supercapacitor are the two electrodes, the separator, and the electrolyte, as shown in Figure 3.5. The electrodes are highly conductive and also possess a large surface area. This is made possible by using metallic current conductors and active materials like carbon. The separator of the two electrodes is essentially a membrane that allows the cross movement of ions but not the electrons. In other words, it prevents electronic conductance. The electrode–separator system is then rolled or folded into a cylindrical or rectangular shape and stacked in a container and impregnated with an electrolyte. Solid-state electrolytes as well as organic or aqueous electrolytes are in use, chosen according to the power requirements.

The voltage at which the supercapacitor is operated is determined by the voltage at which the electrolyte decomposes, which depends on several factors such as the environmental temperature, current intensity, and required lifetime. An EDLC gives a very high capacitance, of the order of thousands of farads, which results from the very small distance that separates the opposite charges at the interfaces between the electrolyte and the electrodes, as well as the large surface area of the electrodes.[11]

3.5.2   Electrodes

As pointed out earlier, the larger capacitance value of a supercapacitor is due to the large surface area of the electrodes, which ensures better contact with the electrolyte and thus increased capacitance. For this purpose, electrochemically inert materials with the highest specific surface area are utilized

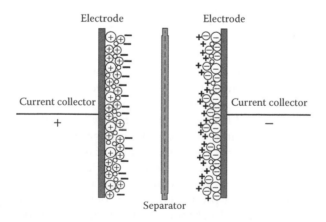

FIGURE 3.5
Schematic representation of a supercapacitor.

to form a double layer with the maximum number of electrolyte ions. The common materials in use are metal oxides, carbon, and graphite, which are characterized by very high surface reactivity. The challenge is the construction of inexpensive electrodes that are chemically and electrically compatible with the electrolyte. In order to be used for applications that demand high energy storage, the electrodes should be constructed with activated carbon (porous) with a very high surface area with appropriate surface and pore geometry. Carbonaceous materials in common use consist of activated carbon fibers, carbon black, active carbon, carbon gel, skeleton carbon, mesocarbon, and microbeads. The best carbon electrodes reported have surfaces with areas as high as 3000 m^2/g of material. The electrode capacitance may reach up to 250 F/g. Carbon powders, or sometimes fibers, are made into a paste and applied on the surface of the metallic current collector. This might cause the development of considerable contact resistance between grains and between the grains and the support. To avoid these problems, some pressure has to be applied, or the carbon powder has to be reinforced with metal fibers or powders, which will increase the electrical conductivity.

Recently, several capacitors, using high–surface area electrodes composed of RuO_2-based composites, have been introduced. Such devices have a capacitance of 200–300 mF at 4–8 V. The Resistance Capacitance (RC) time constant is about 5 ms. The astonishing fact is that they might only be the size of a credit card.[11,12]

3.5.3 Electrolyte

As mentioned earlier, the electrolytes in use can be solid, organic, or aqueous. Quaternary salts dissolved in organic solvents constitute organic electrolytes. The voltage at which they dissociate may be higher than 2.5 V. Aqueous electrolytes are typically KOH or H_2SO_4, with a dissociation voltage of only 1.23 V. The energy density is thus about four times higher for an organic electrolyte when compared with an aqueous one. As evident from Equation 3.3, the energy density varies as the square of the capacitor voltage. Hence, it is desirable for high-energy applications to use a capacitor with an organic electrolyte. Lower electrolyte conductivity will lead to an increase in the internal resistance or ESR. Therefore, care must be taken to ensure that the electrolyte used exhibits high conductivity and adequate electrochemical stability to allow the capacitor to be operated at the highest possible voltages.

Earlier studies have identified tetraethyl ammonium tetrafluoroborate in acetonitrile as the best performing organic electrolyte system for EDLC applications. Depending on the molarity, conductivities up to 60 $mScm^{-1}$ are possible. The market is now buzzing with the introduction of another kind of electrolyte namely ionic liquid, introduced by Covalent Associates, which has many advantages. The ionic liquids are noncorrosive, with typical conductivity around 8 $mScm^{-1}$, and these electrolytes can be used up to a high temperature of 150°C. Blending these electrolytes with acetonitrile is found to enhance the conductivity up to 60 mScm^{-1}.[11,13]

3.5.4 Separator

For the high performance of EDLCs, the need has arisen to look beyond the separators that have been prepared commercially for batteries. Depending on the type of electrolytes, different types of separators are being used. For organic electrolytes, polymer or paper separators are applied. With aqueous electrolytes, glass fiber separators as well as ceramic separators are found to be suitable.

The basic requirements to realize a competitive EDLC are to achieve all the performance characteristics such as high ionic electrolyte conductance, high ionic separator conductance, high electronic separator resistance, high electrode electronic conductance, large electrode surface area, and low separator and electrode thickness.[11–13]

3.6 Charge Storage Mechanism

Though the detailed charge storage mechanism of the supercapacitor is not fully understood, the following theories pertaining to the electrochemical double layer seem to provide adequate interpretation of the charge storage capability of the supercapacitor.

EDLCs are, as explained earlier, constructed from two carbon-based electrodes, an electrolyte, and a separator. An interface is formed at the boundary between any two dissimilar materials or phases. At every interface, an array of charged particles and oriented dipoles are thought to exist. This array is known as an electrical double layer, and Figure 3.6 provides a schematic diagram of a typical EDLC. Like conventional capacitors, EDLCs store charge electrostatically, or non-faradaically, and there is no transfer of charge between the electrode and electrolyte. They utilize an electrochemical double layer of charge to store energy. When a voltage is applied, charge accumulates on the electrodes. Ions in the electrolyte solution diffuse across the separator into the pores of the electrode of opposite charge but there will not be any back movement, resulting in the recombination of the ions. This accumulation of charged ions on the oppositely charged electrode surfaces leads to the formation of a double layer of charge at each electrode, which contributes to an enhanced energy density. Higher energy densities are further facilitated by having a larger surface area for the double layers.

As discussed earlier, there is no charge transfer between the electrode and electrolyte. Hence there are no chemical or composition changes associated with non-faradaic processes. Absence of any chemical change makes the system reversible, with very high cycling stabilities. The EDLCs are generally operated with stable performance characteristics for many charge–discharge cycles, sometimes as many as 10^6 cycles. But electrochemical batteries are generally limited to only about 10^3 cycles. Since they have high cycling stability, EDLCs are used in deep-sea or mountain environments that are inaccessible for manual handling. The basic unit of an EDLC is depicted in Figure 3.7. Mainly three different models are suggested to explain the charge storage mechanism in EDLCs. They are the Helmholtz model, the Gouy–Chapman model, and the Stern model.[2-5]

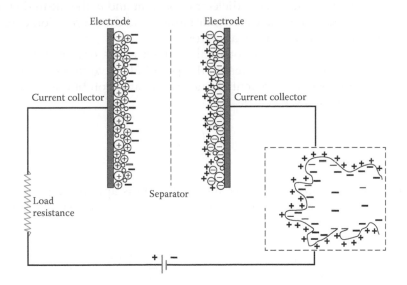

FIGURE 3.6
Schematic diagram of a typical electric double-layer capacitor connected in an electrical circuit.

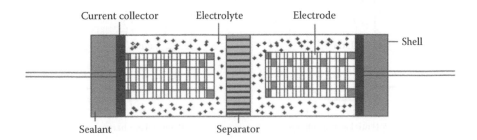

FIGURE 3.7
Basic unit of an electric double-layer capacitor.

3.6.1 Helmholtz Model

According to the Helmholtz model, positive and negative charges stored at the interfaces may be considered as corresponding to a simple parallel plate capacitor. The operating principle of a double-layer capacitor based on the Helmholtz model is illustrated in Figure 3.8.

In general, the relation between the charge per unit area η and the double-layer potential ψ is given by the following equation:

$$\eta = \frac{d}{4\pi\delta} \times \psi \tag{3.7}$$

Here, d is the interface media dielectric constant and δ, the mean distance between the solid, polarizable electrode surface and the average ionic center. The value of δ is generally a few angstroms.

In the Helmholtz model, a potential gradient exists only in the area of the electric double layer. As a result, the potential curve takes the form as shown in Figure 3.8. If the potential under no-charge conditions is ψ_0, then Equation 3.7 can be rewritten as

$$\eta_0 = \frac{d}{4\pi\delta} \times \psi_0 \tag{3.8}$$

On the other hand, if an external electric field is applied to the system, some charges are accumulated at the interfaces. As a result, the value

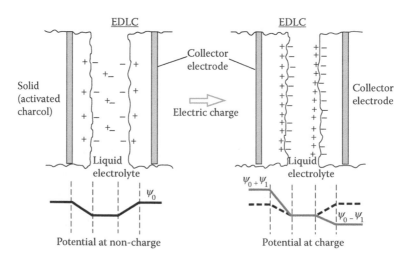

FIGURE 3.8
Helmholtz model of a double-layer capacitor.

of ψ_0 changes to ψ_1 and the charge η_1 gets accumulated, as given by Equation 3.9:

$$\eta_1 = \frac{d}{4\pi\delta} \times (2\psi_1 - \psi_0)$$ (3.9)

From Equations 3.8 and 3.9,

$$\eta_1 = 2\eta_0 \left(\frac{\psi_1}{\psi_0} \right), \quad (\psi_1 \gg \psi_0)$$ (3.10)

It is understood, from Equation 3.10, that the charge equivalent to η_1 can be accumulated by charging with an external electric field. The unit area capacitance, C, is related to the surface charge density, η, and thus to the current density, $i(t)$ where t is the time, via the following equation:

$$C = \frac{d\eta}{d\psi} = \frac{i(t)}{\dfrac{d\psi}{dt}}$$ (3.11)

When a capacitor is under a constant-rate discharge, $i(t) = i$, and

$$C = \frac{i}{\dfrac{d\psi}{dt}}$$ (3.12)

Thus, at a constant current density discharge, one can qualitatively distinguish between battery, supercapacitor, and conventional electrolytic capacitor as shown in Figure 3.9.

The specific capacitance of EDLC is given by

$$\frac{C}{A} = \frac{\varepsilon}{4\pi\delta}$$ (3.13)

Here, C is the capacitance, A is the surface area, ε is the relative dielectric constant of the medium between the two layers (the electrolyte), and δ is the distance between the two layers (the distance from the electrode surface to the center of the ion layer). This approximation is roughly correct for concentrated electrolytic solutions.

The variation in double-layer potential ψ with respect to time is shown in Figure 3.9. It is evident that, the electrolytic capacitor has the least charge storage capacity followed by the supercapacitor having moderate capacity and finally, the battery, which is the best, with apparently infinite charge carrying capacity.[14–17]

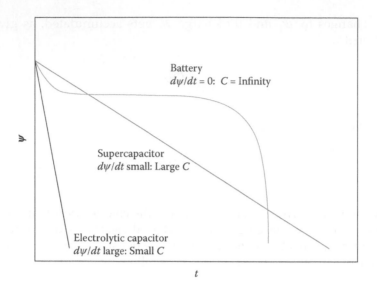

FIGURE 3.9
Variation of double-layer potential ψ with respect to time.

3.6.2 Gouy–Chapman Theory

Gouy–Chapman theory describes the effect of a static surface charge on the membrane potential. A negative surface charge leads to the formation of a double layer, since positive ions in solution tend to balance the negative surface charge. For simplicity, one has to consider only the external surface charge. This is the relevant case for most experiments, where ion substitutions are made only in the external solution.

Consider a qualitative physical description. In the absence of surface charge, the resting potential across the membrane depends on the concentration and membrane permeability of several ions. When external negative surface charge is added, a double layer is formed, and the electric field in this double layer results in a potential difference across it. The electric field in the external double layer is generally in the same direction as the electric field in the membrane. The membrane permeability and the effect of concentration should be balanced so that overall potential differences will not change. The addition of a negative surface charge causes a decrease in the magnitude of the potential difference across the membrane. The usual method for obtaining information about the surface charge is to neutralize it, at least partially, by the addition of polyvalent cations, and to study the voltage-dependent properties for different amounts of neutralization. The neutralization of the negative surface charge may be caused by binding or screening, or by both.

The external surface charge is assumed to be homogeneous and the expression relating to the double-layer potential, the external surface charge density, and the ionic concentrations in the external solution is

$$\sigma_t = \frac{1}{G} \left\{ \sum_{i=1}^{n} C_i \left(\exp\left(\frac{-Z_i FV}{RT} \right) - 1 \right) \right\}^{1/2} \tag{3.14}$$

where
σ_t = negative surface charge density per unit area in the absence of neutralization
G = constant at a given temperature
n = number of ionic species
C_i = concentration of ion i in the external solution
Z_i = valence of ion i
F = faraday constant
V = potential across the double layer
R = gas constant
T = temperature

If σ_t is expressed in units of electronic charges per square nanometer,

$$G = 2.7 \ (\text{nm}^2/\text{electronic charge}) \ (\text{moles}/\text{liter})^{1/2} \tag{3.15}$$

When there is no binding to the surface charges, Equation 3.14 is applicable. In many experiments, the addition of polyvalent cations to the external solution has been used to neutralize surface charges, and thus to probe their properties. When additional polyvalent cations are added to the external solution, the equilibrium constant k for cation binding is given by

$$k = \frac{(MS)}{M_m S} \tag{3.16}$$

where
MS = neutralized site concentration
M_m = polyvalent cation concentration at the membrane
S = free negative surface charge concentration

The relation between M_m, the polyvalent cation concentration at the membrane, and M, the polyvalent cation concentration in the external solution, is given by the Boltzmann factor as follows:

$$M_m = M \exp\left(\frac{-Z_M FV}{RT} \right) \tag{3.17}$$

where Z_M is the valence of polyvalent cation.

The assumption is that each surface charge is either free or completely neutralized. So this is equivalent to assuming that the valence of the polyvalent cations and the valence of the negative surface charge are equal. Then, S_t, the total negative surface charge concentration in the absence of neutralization, is

$$S_t = S + MS \tag{3.18}$$

Equations 3.15 and 3.17 combine to give

$$\frac{S}{S_t} = \frac{1}{(1+kM_m)} \tag{3.19}$$

In the above treatment, S is the negative surface charge concentration per unit volume. If σ is defined as the negative surface charge concentration per unit area, σ and S are proportional, and an equation can be written for σ and σ_t, similar to Equation 3.18 for S and S_t

$$S\frac{\sigma}{\sigma_t} = \frac{1}{(1+kM_m)} \tag{3.20}$$

Combining Equations 3.13, 3.16, and 3.19, one can find a general expression for the charge density

$$\sigma = \frac{1}{G\left(1+kM\exp\left(\frac{-Z_m FV}{RT}\right)\right)} \left\{\sum_{i=1}^{n} C_i\left(\exp\left(\frac{-Z_i FV}{RT}\right)-1\right)\right\}^{1/2} \tag{3.21}$$

It is seen that Equation 3.21 relates the effective surface charge density to the membrane potential V and to the concentration of a neutralizing cation. This concentration appears in the term containing M, which is related to binding, and in the one containing C_i, which is related to screening.

Gouy suggested that interfacial potential at the charged surface could be attributed to the presence of a number of ions of given sign attached to the surface and to an equal number of ions of opposite charge in the solution. Counter ions are not rigidly held, but tend to diffuse into the liquid phase until the counter potential set up by their departure restricts this tendency. The kinetic energy of the counter ions will, in part, affect the thickness of the resulting diffuse double layer. Gouy and, independently, Chapman developed theories of the so-called diffuse double layer, in which the change in concentration of the counter ions near a charged surface is found to follow the Boltzmann distribution:

$$n = n_0 \exp\left(\frac{-ze\psi}{kT}\right)$$ (3.22)

where
n_0 = bulk concentration of counter ions
z = charge on the ion
e = charge on a proton
k = Boltzmann constant.

However, there is an error in this approach, as derivation of this form of the Boltzmann distribution assumes that activity is equal to molar concentration. This may be an acceptable approximation for the bulk solution, but will not be true near a charged surface.

Since there is a diffuse double layer, rather than a rigid double layer, as shown in Figure 3.10, the concern must be with the volume charge density rather than surface charge density when studying the coulombic interactions between charges. The volume charge density for any ion i can be expressed as

$$\rho_i = \sum Z_i e n_i$$ (3.23)

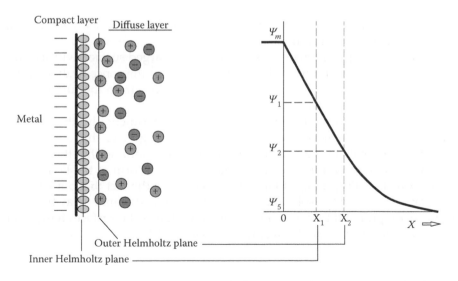

FIGURE 3.10
Helmholtz double layer.

Here the symbols have the already explained meaning. The coulombic interaction between charges can then be expressed by the Poisson equation. For plane surfaces, this can be expressed as

$$\frac{d^2\psi}{dx^2} = \frac{-4\pi\rho}{d} \qquad (3.24)$$

where ψ varies from ψ_0 at the surface to 0 in bulk solution. Thus, one can relate the charge density at any given point to the potential gradient away from the surface.

Combining the Boltzmann distribution with the Poisson equation and integrating under appropriate limits, the electric potential as a function of distance from the surface can be obtained. The thickness of the diffuse double layer is given by

$$\lambda_{\text{double}} = \left[\frac{\varepsilon_r KT}{\left(4\pi e^2 \sum n_{i0} Z_i^2\right)} \right]^{1/2} \qquad (3.25)$$

At room temperature, this expression can be simplified as

$$\lambda_{\text{double}} = 3.3 \times 10^6 \frac{\varepsilon_r}{\left(zc^{1/2}\right)} \qquad (3.26)$$

In other words, the double-layer thickness decreases with increasing valence and ion concentration.

The Gouy–Chapman theory describes a rigid charged surface, with a cloud of oppositely charged ions in the solution and with the concentration of the oppositely charged ions decreasing with the distance from the surface. This is the so-called diffuse double layer.

This theory is still not entirely accurate. Experimentally, the double-layer thickness is generally found to be somewhat greater than calculated. This may be related to the error incorporated by assuming that the activity equals molar concentration when using the desired form of the Boltzmann distribution. Conceptually, it tends to be related to the fact that both anions and cations exist in the solution and with increasing distance away from the surface, the probability that an ion of the same sign as the surface charge will be found within the double layer increases as well.[18–23]

3.6.3 Stern Modification of the Diffuse Double Layer

The Gouy–Chapman theory was developed on a better approximation, compared to the Helmholtz model. The approximation was based on the assumption that ions behave like point charges and also that there are no physical limits for the ions in their approach to the surface. However, strictly speaking, this assumption is not true. Stern, therefore, modified the Gouy–Chapman double-layer theory. His theory states that ions do have a finite size and so cannot approach the surface closer than a few nanometers. The ions of the Gouy–Chapman diffuse double layer are not at the surface, but at some distance δ away from the surface. This distance will usually be taken as the radius of the ion. As a result, the electric potential and the ion concentration of the diffuse part of the layer are low enough to justify treating the ions as point charges.

According to Stern, it is possible that some of the ions are specifically adsorbed by the surface in the δ plane, and this adsorbed layer is known as the Stern layer. Therefore, the potential will drop by $\psi_0 - \psi_\delta$ over the "molecular condenser" (i.e., the Helmholtz plane) and by ψ_δ over the diffuse layer. ψ_δ is termed the zeta (ς) potential. A visual comparison of the amount of counter ions in each of the Stern layer and the diffuse layer is provided in Figure 3.11.

The double layer is formed in order to neutralize the charged surface, which, in turn, brings about an electrokinetic potential between the surface and any point in the mass of the suspending liquid. This voltage difference is of the order of mV and is referred to as the surface potential. The magnitude of the surface potential is related to the surface charge and thickness of the double layer. As one leaves the surface, the potential drops off roughly linearly in the Stern layer and then exponentially through the

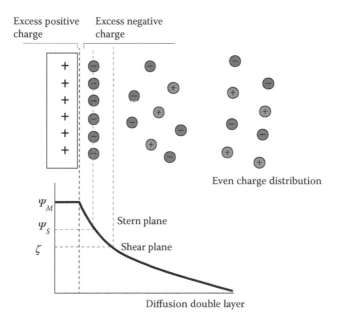

FIGURE 3.11
A comparison of the amount of counter ions in each of the Stern layer and the diffuse layer.

diffuse layer, approaching zero at the imaginary boundary of the double layer. The potential curve is useful because it indicates the strength of the electrical force between particles and the distance at which this force comes into play. A charged particle will move with a fixed velocity in a voltage field. This phenomenon is called electrophoresis. The particle's mobility is related to the dielectric constant and viscosity of the suspending liquid and to the electrical potential at the boundary between the moving particle and the liquid. This boundary is called the shear plane and is usually defined as the point where the Stern layer and the diffuse layer meet. The relationship between zeta potential and surface potential depends on the level of ions in the solution. Figure 3.11 represents the change in charge density through the diffuse layer. The electrical potential at this junction is related to the mobility of the particle and is called the zeta potential. Although zeta potential is an intermediate value, it is sometimes considered to be more significant than surface potential as far as electrostatic repulsion is concerned.[24-26]

Effectively, the capacitance in the EDLC (C_{dl}) can be treated as a combination of capacitances from two regions, the Stern type of compact double-layer capacitance (C_H) and diffusion-region capacitance (C_{diff}). Thus, C_{dl} can be expressed by the following equation:

$$\frac{1}{C_{dl}} = \frac{1}{C_H} + \frac{1}{C_{diff}}$$

(3.27)

3.7   Equivalent Model of an EDLC

A simple model for a double-layer capacitor can be represented by a capacitance (C) with an ESR and an equivalent parallel resistance (EPR) as shown in Figure 3.12. The ESR models power losses that may result from internal heating, which will be of importance during charging and discharging. The EPR models current leakage and influences long-term energy storage. By determining these three parameters, it is possible to develop a first-order approximation of EDLC behavior.

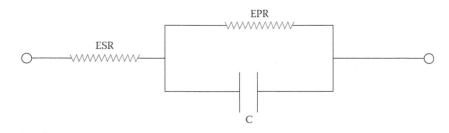

FIGURE 3.12
A simple model for a double-layer capacitor.

3.8 Pseudocapacitance

A different kind of capacitance can arise at certain types of electrodes, called pseudocapacitance, which originates from thermodynamic reasons and is due to charge acceptance (∇q) and a change in potential (∇V). The derivative

$$C = \frac{d(\nabla q)}{(d\nabla V)} \tag{3.28}$$

corresponds to a capacitance, which is referred to as pseudocapacitance. The main difference between pseudocapacitance and EDL capacitance lies in the fact that pseudocapacitance is faradaic (faradaic process is the one that involves the transfer of electrons across the electrode–electrolyte interface) in origin, involving fast and reversible redox reactions between the electrolyte and electroactive species on the electrode surface. The most commonly known active materials showing pseudocapacitance are ruthenium oxide, manganese oxide, vanadium nitride, electrically conducting polymers, such as polyaniline, and oxygen- or nitrogen-containing surface functional groups.

While pseudocapacitance can be higher than EDL capacitance, it suffers from the drawbacks of a low power density (due to poor electrical conductivity) and lack of stability during cycling. A typical example is the pseudocapacitance observed in ruthenium dioxide. This kind of pseudocapacitance can originate when an electrochemical charge transfer process takes place to an extended limit by a finite quantity of reagents or of available surface. Several examples of pseudocapacitance can be cited, but the capacitance function is usually not constant and, in fact, is appreciably dependent on potential or state of charge.

A broad range of significant capacitance values arises for ruthenium dioxide electrodes when the process is surface limited and is proceeded in several one-electron stages, where the pseudocapacitance is almost constant (within 5%) over the full operating voltage range. Some other metal oxides also behave similarly, but only over smaller operating voltage ranges. The ruthenium dioxide pseudocapacitance provides one of the best examples of electrochemical (pseudo) capacitance, because, in addition to the almost constant capacitance over a wide voltage range, its reversibility is excellent, with a cycle life of over several hundred-thousand cycles. Furthermore, the pseudocapacitance can increase the capacitance of an electrochemical capacitor by as much as an order of magnitude over that of the double-layer capacitance. However, as the cost of ruthenium dioxide is comparatively high, electrochemical capacitors are employed in military applications only.[28–30]

Another type of material exhibiting pseudocapacitive behavior, which is highly reversible, is the family of conducting polymers, such as polyaniline or derivatives of polythiophene. These are cheaper than ruthenium dioxide, but are less stable, giving only thousands of cycles (still quite attractive) over a wide voltage range.

3.9   Different Kinds of Supercapacitors: Hybrid Capacitors

Supercapacitors are expected to overcome the relative disadvantages of EDLCs and pseudocapacitors to realize better performance characteristics. Utilizing both faradaic and non-faradaic processes to store charge, hybrid capacitors have achieved energy and power densities greater than EDLCs without the sacrifices in cycling stability and affordability that have limited the success of pseudocapacitors. Research has been focused on three different types of hybrid capacitors, distinguished by their electrode configuration. They are the composite, asymmetric, and battery types of hybrid capacitors.

In composite electrodes, carbon-based materials are integrated with either conducting polymers or metal oxide materials, and they incorporate both physical and chemical charge storage mechanisms together in a single electrode. The carbon-based materials facilitate a capacitive double layer of charge and also provide a high–surface area backbone that increases the contact between the deposited pseudocapacitive materials, and the electrolyte. The pseudocapacitive materials further increase the capacitance of the composite electrode through faradaic reactions.

Asymmetric hybrids combine faradaic and non-faradaic processes by coupling an EDLC electrode with a pseudocapacitor electrode. In particular, the coupling of an activated carbon-negative electrode with a conducting polymer-positive electrode has received a great deal of attention. The lack of an efficient, negatively charged conducting polymer material has limited the success of conducting-polymer–based pseudocapacitors. Asymmetric hybrid capacitors that couple these two electrodes mitigate the extent of this trade-off to achieve higher energy and power densities than comparable EDLCs. Also, they have better cycling stability than comparable pseudocapacitors.

Like asymmetric hybrids, battery-type hybrids couple two different electrodes. However, battery-type hybrids are unique in coupling a supercapacitor electrode with a battery electrode. This specialized configuration reflects the demand for higher energy supercapacitors and higher power batteries, combining the energy characteristics of batteries with the power, cycle life, and recharging characteristics of supercapacitors.

Because of the difference in the mechanisms of storing charges between the double-layer capacitors and the conventional capacitors, traditional capacitor models are inadequate for electrochemical capacitors. A number of models currently exist that apply to the operation of double-layer capacitors.[31–37]

3.10   Electrode Materials for EDLCs and Supercapacitors

3.10.1   Activated Carbons

Activated carbons are less expensive and possess a higher surface area than other carbon-based materials, and activated carbon is the most commonly used electrode material in EDLCs. Activated carbons have a complex porous structure composed of differently sized micropores (<20 Å wide), mesopores (20–500 Å), and macropores (>500 Å) to achieve high surface areas. Although capacitance is directly proportional to the surface area, for activated carbons, all of the high surface area does not contribute to the capacitance of the device.

3.10.2   Carbon Aerogels

There is much interest in using carbon aerogels as electrode materials for EDLCs. Carbon aerogels are formed from a continuous network of conductive carbon nanoparticles with intermingled mesopores. Due to this continuous structure and its ability to bond chemically to the current collector, carbon aerogels do not require the application of an additional adhesive binding agent. As binderless electrodes, carbon aerogels have been shown to have a lower ESR than activated carbons.

3.10.3   Carbon Nanotubes

Recent research trends are based on the use of carbon nanotubes as an EDLC electrode material. Electrodes made from carbon nanotubes are grown as an entangled mat of carbon nanotubes, with an open and accessible network of mesopores. Unlike other carbon-based electrodes, the mesopores in carbon nanotube electrodes are interconnected, allowing a continuous charge distribution that uses almost all of the available surface area. Thus, the surface area is utilized more efficiently to achieve capacitance values comparable to those in activated-carbon–based supercapacitors, even though carbon nanotube electrodes have a modest surface area compared with activated carbon electrodes. Because the electrolyte ions can more easily diffuse into the mesoporous network, carbon nanotube electrodes also have a lower ESR than activated carbon electrodes.

3.10.4   Conducting Polymers

Conducting polymers have a relatively high capacitance and electrical conductivity and they also have lower ESR and cost compared to carbon-based electrode materials. In particular, the n/p-type polymer configuration, with one negatively charged (n-doped) and the other positively charged (p-doped) conducting polymer electrode, has the highest power density and potential energy. Pseudocapacitors are not utilized to their full potential because of the lack of efficient, n-doped conducting polymer materials. Additionally, it is believed that the mechanical stress on conducting polymers during reduction–oxidation reactions limits the stability of these pseudocapacitors through many charge–discharge cycles.

3.10.5   Composite Materials

Composite electrodes are formed by the combination of carbon-based materials with either conducting polymers or metal oxide materials as explained earlier.[38–41]

3.11   Comparison

The specific energy versus specific power graph, for different energy storage devices is shown in Figure 3.13, and the interesting thing is that supercapacitors cover the maximum area in the diagram.[2]

Various parameters of a battery, electrostatic capacitor, and electrochemical double-layer capacitor are compared in Table 3.1. It can be inferred that supercapacitors or electrochemical double-layer capacitors hold healthy characteristics and dominate batteries and electrostatic capacitors in most of the parameters.[42–47]

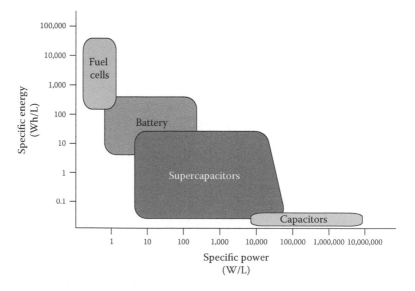

FIGURE 3.13
Specific energy versus specific power graph.

TABLE 3.1

Comparison of Parameters of a Battery, Electrostatic Capacitor, and Electrochemical Double-Layer Capacitor

	Battery	Electrostatic Capacitor	Electrochemical Capacitor
Discharge time	0.3–3 h	10^{-3} to 10^{-6} s	0.3–30 s
Charge time	1–5 h	10^{-3} to 10^{-6} s	0.3–30 s
Energy density (Wh/kg)	10–100	<0.1	1–10
Specific power (W/kg)	50–200	>10,000	≈1000
Charge–discharge efficiency	0.7–0.85	≈1	0.85–0.98
Cycle life	500–2000	>500,000	>100,000

3.12 Applications

The supercapacitor is still a young technology that is yet to experience widespread implementation. It does, however, enjoy a great amount of attention with regards to its potential applications in a number of areas.

A traditionally high ESR has previously limited EDLCs to memory backup applications, and they have been used in such settings for many years. Recent reductions in ESR have improved the power capabilities of supercapacitors, and they are now well suited to pulsed-current applications such as mobile phones and electrical actuators. They can also perform a load-leveling function when used in combination with batteries, providing peak power in devices, such as laptops, and reducing power demands on the battery and, therefore, extending the battery's lifetime. They can be used in a similar manner in electrical vehicles (EVs), providing power for acceleration and allowing a primary power source, such as a fuel cell, to supply the average power. When used in EVs, supercapacitors also allow for energy to be recuperated during braking, improving the efficiency of the vehicle.

Supercapacitors can also be used on their own to provide the energy needed by power quality systems that ensure reliable and disturbance-free power distribution. Supercapacitors then supply the energy needed to inject power into the distribution line and thus compensate for any voltage fluctuations. They can also be used to design systems that grant adjustable-speed drives the ability to ride through temporary power supply disturbances. Such applications are vital in industrial settings and can prevent material and financial losses that could occur due to machine downtime.

3.13   Advantages and Disadvantages of Supercapacitors

The advantages and certain disadvantages of supercapacitors are listed in Table 3.2.

3.14   Concluding Remarks

With the advent of nanoscience and nanotechnology, many novel ideas have been introduced to revolutionize supercapacitor design and performance. In 2007,[48] a supercapacitor architecture employing vertically aligned single-walled carbon nanotubes (SWNTs) as electrodes and a room temperature ionic liquid as the electrolyte was demonstrated, with all the components integrated on cellulose paper. The large surface area associated with the nanotube electrodes and the much reduced spacing between them have resulted in much higher values of capacitance, of the order of several thousands of farads. Since then, there have been a flow of innovative ideas to make the so-called paper supercapacitors smarter. In one of the recent developments,[49] a modified paper supercapacitor has been realized by printing SWNTs onto treated sheets of paper. These supercapacitors are found to have capacitance around 3 F/g and excellent cycling stability up to 2500 cycles without capacity loss. Paper-based, all solid-state, flexible supercapacitors are of very recent origin. They can be charged by solar cells and then

TABLE 3.2

The Various Advantages and Certain Disadvantages of Supercapacitors

Advantages	Disadvantages
• Device voltage is determined by the circuit application, not limited by the device chemistry. • Very high device voltages are possible (but there is a trade-off with capacity). • High power is available. • High power density is possible. • Simple charging methods can be adopted. No special charging or voltage detection circuits are required. • Can be charged and discharged in seconds, much faster than batteries. • No chemical reactions. • Cannot be overcharged. • Long cycle life of more than 500,000 cycles at 100% depth of discharge. • Long calendar life of 10–20 years. • Low impedance	• Linear discharge voltage characteristics prevent the use of all the available energy in some applications. • Power is available only for a very short duration. • Low energy density (6 Wh/kg) · • Cell balancing required for series chains. • High self-discharge rate, which is much higher than that of batteries.

discharged to power many devices, demonstrating their efficient energy management in self-powered nanosystems.[50] Another exciting event is the practical realization of paintable supercapacitors.[51] In this approach, the supercapacitor components are engineered into paint formulations and simple spray painting techniques can be used to fabricate the supercapacitors on a wide variety of materials, including glass, glazed ceramic tiles, flexible polymer sheets, and stainless steel, without the requirement of any surface conditioning. By using the spray painting technique, it is possible to achieve maximum flexibility in surface forms and device geometries. This technique can equally well be used to fabricate rechargeable Li ion cells and solar cells on any desired surfaces. The possibility of assembling such devices on any arbitrary surface will have a significant impact on the design, implementation, and integration of energy-capture storage devices.

References

1. Simon P, Gogotsi Y. Materials for electrochemical capacitors. *Nat Mater* 2008; 7: 845–854.
2. Kötz R, Carlen M. Principles and applications of electrochemical capacitor. *Electrochim Acta* 2000; 45: 2483–2498.
3. Jayalakshmi M, Balasubramanian K. Simple capacitors to supercapacitors—An overview. *Int J Electrochem Sci* 2008; 3: 1196–1217.
4. Halper MS, Ellenbogen JC. *Supercapacitors: A Brief Overview*. Virginia, USA: McLean, 2006.
5. Conway BE. *Electrochemical Supercapacitors: Scientific Fundamentals and Technological Applications*. New York: Kluwer-Plenum, 1999.
6. Burke A. Ultracapacitors: Why, how, and where is the technology? *J Power Sources* 2000; 91: 37–50.
7. Chu A, Braatz P. Comparison of commercial supercapacitors and high power lithium-ion batteries for power-assist applications in hybrid electric vehicles I. Initial characterization. *J Power Sources* 2002; 112: 236–246.
8. Georgiev AM. The electrolytic capacitor. *Nature* 1948; 162: 911.
9. Miller JR, Simon P. *Fundamentals of Electrochemical Capacitor Design and Operation, The Electrochemical Society Interface*. Spring, 2008.
10. Conway BE. *Electrochemical Supercapacitors: Scientific Fundamentals and Technological Applications*. New York: Kluwer Academic/Plenum Publishers, 1999.
11. Schneuwly A, Gallay R. Properties and applications of supercapacitors from the state-of-the-art to future trends. *Proc PCIM* 2000.
12. Beliakov A. *ELIT Company, 9th Seminar ECDL*. Deerfield Beach, FL, 1999.
13. Ue D, Ida K, Mori S. Electrochemical properties of organic liquid electrolytes based on quaternary onium salts for electrical double-layer capacitors. *J Electrochem Soc* 1994; 141: 2989.
14. Bullard GL, Sierra-Alcazar HB, Lee HL, Morris JL. Operating principles of the ultra capacitor. *IEEE Trans Magn* 1989; 25: 1.

15. Byker HJ, Vaaler LE. High-discharge-capacity bipolar and capacitive batteries. Report R-6188, U.S. Army Missile Command, Redstone Arsenal, AL 1985.
16. Lee HL, Bullard GL, Mason GE, Kern K. Improved pulse powerll sources with high-energy density capacitor. *IEEE, 4th Symposium on Electromagnetic Launch Technology*, Austin, TX 1988: 12–14.
17. O'M Bockris J, Reddy AKN. *Modern Electrochemistry*. New York: Plenum Press, 1970.
18. Neal C, Cooper DM. Extended version of Gouy-Chapman electrostatic theory as applied to the exchange behavior of clay in natural waters. *Clay Clay Miner* 1983; 31: 367–376.
19. Kinraide TB. Use of a Gouy-Chapman-Stern model for membrane-surface electrical potential to interpret some features of mineral rhizotoxicity. *Plant Physiol* 1994; 106: 1583–1592.
20. Torrie GM, Valleau JP. Electrical double layers. 4. Limitations of the Gouy-Chapman theory. *J Phys Chem* 1982; 86: 3251–3257.
21. Valleau JP, Torrie GM. The electrical double layer. III. Modified Gouy-Chapman theory with unequal ion sizes. *J Chem Phys* 1982; 76: 4623–4630.
22. Bolt GH. Analysis of the validity of the Gouy-Chapman theory of the electric double layer. *J Colloid Sci Imp U Tok* 1955; 10: 206–218.
23. Oldham KB. A Gouy–Chapman–Stern model of the double layer at a (metal)/(ionic liquid) interface. *J Electroanal Chem* 2008; 613: 131–138.
24. Prigogine I, Rice SA, Carnie SL, Torrie GM. The statistical mechanics of the electrical double layer. *Adv Chem Phys* 2007; 56: 141–253.
25. Mangelsdorf CS, White LR. Effects of stern-layer conductance on electrokinetic transport properties of colloidal particles. *J Chem Soc Faraday Trans* 1990; 86: 2859–2870.
26. Zukoski CF, Saville DA. The interpretation of electrokinetic measurements using a dynamic model of the Stern layer: I. The dynamic model. *J Colloid Interface Sci* 1986; 114: 32–44.
27. Miller JR, Simon P. Fundamentals of electrochemical capacitor design and operation. *Electrochemical Soc Interface*, Spring, 2008.
28. Burke A. Ultracapacitors: Why, how, and where is the technology? *J Power Sources* 2000; 91: 37–50.
29. Beliakov AL, Brintsev AM. Development and application of combined capacitors: Double electric layer—Pseudocapacity. *Proceedings of the 7th International Seminar on Double-layer Capacitors and Similar Energy Storage Devices*, Deerfield Beach, FL, 1997.
30. Conway BE, Birss V, Wojtowicz J. The role and utilization of pseudocapacitance for energy storage by supercapacitors. *J Power Sources* 1997; 66: 1–14.
31. Lin C, Ritter JA, Popov BN, et al. Correlation of double-layer capacitance with the pore structure of sol-gel derived carbon xerogels. *J Electrochem Soc* 1999; 146: 3639–3643.
32. Celzard A, Collas F, Mareche JF, et al. Porous electrodes-based double-layer supercapacitors: Pore structure versus series resistance. *J Power Sources* 2002; 108: 153–162.
33. Laforgue A, Simon P, Fauvarque JF, et al. Activated carbon/conducting polymer hybrid supercapacitors. *J Electrochem Soc* 2003; 150: A645–A651.
34. Mastragostino M, Arbizzani C, Soavi F, et al. Conducting polymers as electrode materials in supercapacitors. *Solid State Ionics* 2002; 148: 493–498.

35. Li HQ, Cheng L, Xia Y, et al. A hybrid electrochemical supercapacitor based on a 5 V Li-ion battery cathode and active carbon. *Electrochem Solid State Lett* 2005; 8: A433–A436.
36. Wang X, Zheng JP. The optimal energy density of electrochemical capacitors using two different electrodes. *J Electrochem Soc* 2004; 151: A1683–A1689.
37. Du Pasquier A, Plitz I, Menocal S, et al. A comparative study of Li-ion battery, supercapacitor and nonaqueous asymmetric hybrid devices for automotive applications. *J Power Sources* 2003; 115: 171–178.
38. Qu DY, Shi H. Studies of activated carbons used in double-layer capacitors. *J Power Sources* 1998; 74: 99–107.
39. Gamby J, Taberna PL, Simon P, et al. Studies and characterisations of various activated carbons used for carbon/carbon supercapacitors. *J Power Sources* 2001; 101: 109–116.
40. Shi H. Activated carbons and double layer capacitance. *Electrochim Acta 1996*; 41: 1633–1639.
41. Wang J, Zhang SQ, Guo YZ, et al. Morphological effects on the electrical and electrochemical properties of carbon aerogels. *J Electrochem Soc* 2001; 148: D75–D77.
42. Zhang Y, Feng H, Wu X, et al. Progress of electrochemical capacitor electrode materials: A review. *Int J Hydrogen Energ* 2009; 34: 4889–4899.
43. Lota G, Centeno TA, Frackowiak E, Stoeckli F. Improvement of the structural and chemical properties of a commercial activated carbon for its application in electrochemical capacitors. *Electrochim Acta* 2008; 53: 2210–2216.
44. Simon P, Gogotsi Y. Materials for electrochemical capacitors. *Nat Mater* 2008; 7: 845–854.
45. Erdinc O, Vural B, Uzunoglu M, Ates Y. Modeling and analysis of an FC/UC hybrid vehicular power system using a waveletfuzzy logic based load sharing and control algorithm. *Int J Hydrogen Energ* 2008; 10: 39.
46. Suppes GJ. Plug-in hybrid with fuel cell battery charger. *Int J Hydrogen Energ* 2005; 30: 113–121.
47. Pandolfo AG, Hollenkamp AF. Carbon properties and their role in supercapacitors. *J Power Sources* 2006; 157: 11–27.
48. Pushparaj VL, Shaijumon MM, Kumar A, et al. Flexible energy storage devices based on nanocomposite paper. *Proc Natl Acad Sci USA* 2004; 104: 13574–13577.
49. Hu L, Wu H, Cui Y. Printed energy storage devices by integration of electrodes and separators into single sheets of paper. *Appl Phys Lett* 2010; 96: 183502.
50. Yuan L, Xiao X, Ding T, et al. Paper-based supercapacitors for self-powered nanosystems. *Angew Chem Int Ed* 2012; 51: 4934–4938.
51. Singh N, Galande C, Miranda A, et al. Paintable battery. *Sci Rep* 2012; 2: 481.

4

Measurement Techniques for Performance Evaluation of Supercapacitor Materials and Systems

Heather Andreas

CONTENTS

4.1   Introduction

In order for a material to be considered a supercapacitor, it should have electrochemical properties similar to that of a typical capacitor. This chapter describes the common measurement techniques used to evaluate the performance of a capacitive material or supercapacitor system. These techniques include cyclic voltammetry, galvanostatic charge/discharge, electrochemical impedance spectroscopy (EIS), self-discharge measurements, and others. A brief analysis of the typical experimental profiles expected for each technique is provided, as well as a description of the method by which key supercapacitor descriptors can be calculated.

Although this book focuses on the pseudocapacitive metal oxide supercapacitor systems, in practice, many of these metal oxides are deposited on high–surface area carbon electrodes in order to minimize cost and weight. The charge storage in these carbon electrodes is primarily double-layer storage; therefore, a general description of simplified double-layer models is provided herein. These high–surface area carbon electrodes may also introduce resistance limitations in the material because of the, sometimes significant, electrolyte resistance inside the pores. As many of the key supercapacitor characteristics depend on the porosity of the electrode, a description of the pore effect is provided. Where appropriate, the resistive considerations of ion migration through pseudocapacitive films is also considered with respect to the influence this resistance will have on the expected experimental profiles obtained by the techniques outlined in this chapter.

4.2   Definitions and Characteristics of Key Descriptors

This section defines some of the key descriptors used in evaluating supercapacitors and the key relationships between these descriptors. Important considerations that influence each descriptor and supercapacitor performance are also highlighted.

4.2.1 Voltage/Potential

Potential is the energy required to move a unit charge through an electric field, or in the case of an electrode–electrolyte interface, across the interface. Often, the terms "potential" and "voltage" are used interchangeably. More precisely, "potential" is used when the measurement is made against an electrode of known constant value such as a reference electrode. "Voltage" is used when recording the difference in voltage between two nonreference electrodes (e.g., the positive and negative electrodes of a full supercapacitor). In this chapter, voltage and potential are both denoted with the symbol V. The unit for voltage and potential is volts (V).

4.2.2 Stable Voltage Window

One of the key characteristics of a supercapacitor electrode or electrolyte material is its stable voltage window. The stable voltage window is the potential over which all the materials in the supercapacitor can be used without degradation (i.e., without irreversible and damaging changes to either electrodes or electrolyte, such as electrode dissolution or irreversible oxidation/reduction, electrolyte decomposition, significant volume changes leading to electrical disconnection of material, etc.). In practice, the voltage window used is often smaller than the stable voltage window to ensure material stability in the event of accidental overcharge. Additionally, for some pseudocapacitive materials with only small potential windows for the pseudocapacitive reaction, the practical voltage window may be limited to that over which only the pseudocapacitive reaction occurs, as this is the range in which charge storage occurs. Thus, the practical window can be much smaller than the stable window for these materials.

Together, the practical, stable voltage windows of the materials in a supercapacitor define the voltage of the supercapacitor as a whole. Since energy is dependent on V^2 (see Equation 4.4), the voltage of the supercapacitor is extremely important in defining the energy that can be stored in a supercapacitor. Additionally, overcharging a supercapacitor (applying potentials outside of the material's stable voltage window) can lead to a decrease in performance over time,[1] thereby reducing the life cycle of the system. Overcharging of supercapacitor materials can also lead to enhanced self-discharge rates.[2]

4.2.3 Current

Current (I) is the amount of charge [Q, in coulombs (C)] transferred per unit time [t, in seconds (s)]. The unit for current is amperes (A).

4.2.4 Resistance

The resistance [R, in ohm (Ω)] of a material to the movement of charge results in a loss of voltage, and therefore energy per unit charge, based on Ohm's law ($V = IR$). Often, the voltage lost through a resistive component is termed the "ohmic" or "iR" drop, where "i" denotes a current density (A/cm^2). Two types of resistance will be described in this chapter: charge transfer resistance and solution resistance. Charge transfer resistance (R_{ct}) is the resistance to the transfer of charge across an electrode–electrolyte interface during a faradaic reaction. Solution resistance (R_s) is the resistance of the electrolyte to ion migration resulting in a loss of energy, often as heat.

4.2.5 Capacity

Capacity is the total amount of charge stored in a fully charged system and therefore has the same unit as charge, that is, coulombs.

4.2.6 Capacitance

Capacitance [C, in farads (F)] is the amount of charged stored on the capacitor at a particular voltage:

$$C = \frac{Q}{V} = \frac{It}{V} \tag{4.1}$$

Since supercapacitors have two capacitive electrodes in series, the overall capacitance of the supercapacitor is related to the capacitance of the two electrodes by

$$\frac{1}{C_t} = \frac{1}{C_{electrode\,1}} + \frac{1}{C_{electrode\,2}} \tag{4.2}$$

The relationship in Equation 4.2 highlights one of the key parameters of supercapacitors, that is, the overall system capacitance is often governed by the smaller capacitance of the two electrodes, or if two electrodes of equal capacitance are used, the overall capacitance of the system is only half of the capacitance of each electrode.[3]

4.2.7 Coulombic Efficiency

Coulombic efficiency (CE), or round-trip CE, is defined differently by different authors. It is usually defined as the percentage ratio of the charge removed from the electrode surface during discharge (Q_{dis}) to that placed on the electrode surface during charge (Q_{ch})

$$CE = \frac{Q_{dis}}{Q_{ch}} \times 100\% \tag{4.3}$$

or as the percentage ratio of cathodic charge (Q_c) to anodic charge (Q_a). Other authors define CE as the ratio of anodic charge to cathodic charge.[1] Thus, this parameter and the method used for its calculation must be defined by authors.

It should be noted that a CE of 100% does not indicate electrochemical reversibility (which relates specifically to the rapid rate of the reaction and not whether the reverse reaction can occur) and is not an indicator of capacitance.[1] CEs of less than 100% are indicative of a faradaic reaction occurring that is not reversible in the potential window of the supercapacitor; for instance, it may indicate a reaction such as electrolyte decomposition or reaction of an impurity. In many situations, a CE <100% suggests degradation of the system, resulting in a shorter life cycle, or a parasitic side reaction, leading to higher self-discharge.

4.2.8   Time Constant

The time constant is given by RC, where R is the resistance to charge transfer in the supercapacitor system (i.e., limits how quickly the charge or ions can move in the supercapacitor) and C is the capacitance (the amount of charge that is stored at the voltage applied to the supercapacitor). Thus, the time constant provides a measure of the amount of charge available in the system and the rate at which the charge can be delivered. Since power (as described in Section 4.2.13) is dependent on the rate of charge delivery, the time constant is a key characteristic for supercapacitor performance.

4.2.9   State of Charge

Essentially, state of charge (SOC) is a descriptor of how charged the supercapacitor is at any given point. Since the voltage of a supercapacitor drops during discharge (and increases during charging), the voltage can be used as a measure of SOC. This is contrary to what is seen in many types of batteries, where the voltage is essentially constant until just before the battery is fully discharged.

4.2.10   Electrochemical Reversibility

Electrochemical reversibility refers to faradaic reactions where the rate of the electron transfer across the interface does not limit the rate of the reaction. Reactions that are pseudocapacitive are electrochemically reversible, and since the rate of the reaction is not limited by activation (the rate of electron transfer), these reactions are at equilibrium and they are governed by a Nernst-type equation relating the degree of the reaction to the potential.[4] For pseudocapacitive reactions based on electrochemical reversibility, at high rates of charge or discharge (i.e., fast sweep rates in cyclic voltammetry, high currents in galvanostatic charge/discharge, or high powers in constant power discharge), the electrochemical experimentation may become too fast for the reaction, resulting in irreversibility and a loss of pseudocapacitive energy density.

4.2.11 Energy

The energy stored [E, in joules (J) or watt-hours (Wh)] depends on the voltage at which the charge may be added to or removed from the system. In supercapacitors, the voltage drops with discharging, which results in a fundamental difference between batteries and supercapacitors, namely that the energy of a supercapacitor is half of that of a battery (where the voltage is essentially constant with discharge).[4] The equations used for calculating energy are

$$E = \int V dQ = ItV = \frac{1}{2}CV^2 \tag{4.4}$$

As the voltage drops during discharging, it leads to a characteristic of supercapacitors that most of their energy is delivered near the supercapacitor's maximum voltage. For a system with constant capacitance over the voltage window (e.g., a pure double-layer supercapacitor, or a supercapacitor with pseudocapacitance that mimics a double-layer supercapacitor), 75% of the energy is dropped in the first half of the voltage window. Thus, often supercapacitors are discharged to only half of their maximum voltage.

Any resistive components of the system (such as charge transfer resistance or solution resistance) will result in a lower energy than that predicted from the system's maximum voltage (V_{max}). With ohmic polarization[1] due to solution resistance, Equation 4.4 is modified to:

$$E = \frac{1}{2}C[V_{max} - iR_s]^2 \tag{4.5}$$

Variables that increase the resistance in the system, such as low electrolyte concentration and low temperatures, will increase the energy lost. This ohmic effect is less significant in electrolytes with high conductivity, such as aqueous electrolytes, particularly acidic or alkaline electrolytes. The higher solution resistance of ionic liquid and organic electrolytes is often offset by their larger stable potential windows (larger V_{max}).[5] Thus, while their higher resistances do result in a more dissipative loss of energy, this is more than offset by their larger potential windows, and their energies are larger than those of aqueous systems.[5]

Other polarization in the system, such as activation polarization for faradaic reactions or concentration polarization caused by depletion of an active species near the electrode surface, will lead to a similar dissipation of energy, where iR_s in Equation 4.5 is replaced by the activation polarization η giving[2]

$$E = \frac{1}{2}Q[V - \eta]^2 \tag{4.6}$$

Clearly, any loss of potential through the system can have a significant negative impact on the system's energy because of the squared relationship between potential and energy. Indeed, at high rates, it is often the dissipative losses due to cell resistance or reaction polarization that limit the energy (and power) characteristics of a supercapacitor.[2] The cumulative solution resistance down the pores in highly porous electrodes can be quite significant. Upon application of a voltage, the current decreases down the pore while the cumulative resistance increases. Therefore, there is no single iR value (nor a single time constant) for a porous system, but rather a range of ohmic loss values varying down the pore. The energy lost through pore resistance is then dependent on the penetration depth of the voltage into the pore.[2] Similar resistance considerations are possible for the resistance to ion migration through pseudocapacitive films; however, typically pseudocapacitive films are synthesized in such a way to minimize the distance of ion migration and therefore minimize these resistive considerations.[6-8]

4.2.12 Energy Efficiency

Energy efficiency (EE) is, again, defined differently by different authors; typically, it is defined as the percentage ratio of energy provided during discharge (E_{dis}) to energy required during charging (E_{ch}):

$$EE = \frac{E_{dis}}{E_{ch}} \times 100\% \tag{4.7}$$

With three-electrode experiments EE may also be defined as the percentage ratio of anodic energy to cathodic energy[1] or vice versa.

In reality, EE will never be 100%, since there will always be some dissipative energy losses due to cell resistance, as described above.[1] Highly porous electrodes (with high resistance due to narrow, long pores or those containing bottlenecks), low electrolyte concentrations, and low temperatures are all expected to reduce the energy and EE of a system.

Pseudocapacitive electrodes may also experience some voltage loss due to activation polarization.[1] Also, as many pseudocapacitive reactions require participation of an electrolyte species (such as H^+), at high rates or when the mass transfer of electrolyte species is limited by the film, these reactions may also experience concentration polarization.[1] Both of these polarizations result in a lower potential during discharge and a higher potential during charging, and therefore a lower E_{ch} and higher E_{dis} and consequently a lower EE.[1]

From Equations 4.4 and 4.7, it can be seen that any electrochemical irreversibility in the system, which results in charge being placed on a supercapacitor at a high voltage (giving a high E_{ch}) but removed at a low voltage (giving a low E_{dis}), will dramatically decrease the EE of the cell.[4] This highlights the importance of the mirror-image requirement in cyclic voltammogram (CV). CVs, which are mirror images about the zero-current line, have the charge being placed on the surface and removed from the surface at the same voltage, meaning $E_{ch} = E_{dis}$ and, therefore, result in a high EE.

Even highly electrochemically reversible reactions with mirror-image CVs may become rate limited if the sweep rate applied is approximately the same as the rate constant for the reaction.[4] In these situations, the peaks in a CV that appeared to be a mirror image at low rates begin to separate at higher rates and eventually give the characteristic electrochemically irreversible shape, where the current around the equilibrium potential is governed by the Butler–Volmer equation.

4.2.13 Power

Power [P, in joules per second (J/s) or watt (W)] is the amount of energy delivered in a particular time and can be calculated in a number of ways:

$$P = \frac{E}{t} = \frac{QV}{t} = IV = I^2R = \frac{V^2}{4R_s} \tag{4.8}$$

Power is one of the most important characteristic calculated for a supercapacitor because the fast kinetics of the double-layer charge or pseudocapacitive reactions allow for these systems to be used in high-power applications. In practical supercapacitors and supercapacitor materials, there are some rate considerations that impact the power. The rate limitations most often stem from resistive effects (see Section 4.2.2 on Resistance), such as resistance to ion migration in pseudocapacitive films or resistance of the electrolyte in the pores of highly porous electrodes (see Section 4.3.2). In order to maximize the power of a system, factors that increase the resistance (such as low electrolyte concentration, highly porous electrodes, and low temperature) must be minimized.

4.3 Theoretical Treatment and Modeling of the Double Layer at Electrode-Electrolyte Interfaces

While this book is focused primarily on metal-oxide supercapacitors, where the charge storage is primarily through pseudocapacitive faradaic reactions, a brief description is provided here about the double layer formed at the electrode–electrolyte boundary. The double layer forms at all electrode–electrolyte boundaries and is responsible for storing some charge on the capacitor electrode; however, for primarily pseudocapacitive systems, the amount of double-layer charge storage is small (5%–10%)[4] relative to the pseudocapacitive charge storage.

As charge is placed on the surface of an electrode in contact with an electrolyte, an excess of oppositely charged ions will accumulate in the electrolyte near the interface to balance the surface charge. The layer of charge on the surface plus the balancing layer of ionic charge in the electrolyte is termed the "double layer."

The structure of the double layer at an electrode–electrolyte boundary is quite complex and has been covered in a number of publications, including Chapter 6 of the book by Conway.[4] As modeled by Grahame,[9] the so-called "double layer" is in fact composed of three layers in solution plus a layer of charge on the electrode surface. Just outside of the electrode surface reside

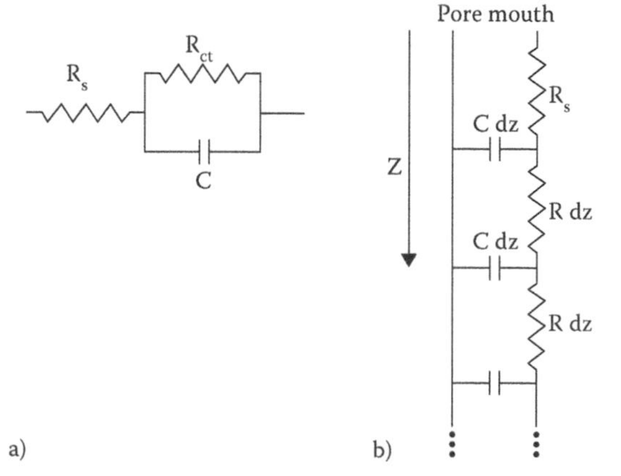

FIGURE 4.1
Some equivalent circuit models used in supercapacitor systems: a) modified Randles circuit and b) transmission line circuit. (Adapted from de Levie, R., *Electrochim. Acta*, 8, 751–780, 1963.)

anions that have lost their hydration shell and are directly, and specifically, adsorbed on the electrode surface. These ions are at the distance of closest approach, called the "inner Helmholtz layer." These two layers of separated charge (one on the surface and one in solution) can be modeled with a parallel plate capacitor model, and as expected from this model, there is a linear drop in potential (or charge) between the surface and inner Helmholtz layer. The cations and anions that have not lost their hydration shell may then sit one or two water layers away from the surface, at the "outer Helmholtz layer." Again, the potential (or charge) drops linearly between the inner and outer Helmholtz layers. Additionally, there is the Gouy–Chapman layer, where there are some ions that are further away from the surface (due to thermal buffeting) but are still attracted electrostatically to the surface. The potential drop in the Gouy–Chapman layer is exponential as described by the Poisson–Boltzmann equation, and is similar to that for the Debye–Hückel model of charge density around an ion; the Gouy–Chapman model predated the Debye–Hückel model by approximately a decade.

Even the Grahame model described earlier is an oversimplification of the double-layer region. Modeling (and experimentation) has shown that the double layer is not one ionic layer thick, but rather has a number of ion layers required to balance the charge on the electrode surface.[10] The first layer of ions overcompensates the surface charge (called overscreening). This extra electrolyte charge is then balanced (and slightly overcompensated) by ions of the opposite charge in the next layer of the electrolyte, and so on,[10,11] making the double-layer model more complex.

Although the double-layer models described before are already very simplified models of the complex region near a charged electrode–electrolyte interface, when considering the interface in supercapacitors, the double layer is often simplified even further to a pure Helmholtz model, where there is a plane of charge on the surface and a plane of ions balancing that charge in solution at some point from the surface. This simplification is justified by the high concentrations of ions used in supercapacitor systems, and relies on the Stern model, which suggests that at high-ion concentration the majority of charge-balancing ions sit on the Helmholtz plane, rather than in the Gouy–Chapman layer.[4,10] The Helmholtz model, with its two planes of balancing charge, and linear potential drop between them, can be modeled electronically in the same way as a parallel-plate capacitor. The capacitance (C) of the electrode–electrolyte interface is then related to the dielectric constant of the material between the layers (ε) (in this case, one or two layers of the electrolyte), the vacuum permittivity (ε^o), the area of the electrode (A), and the distance (d) between the surface and the ionic-charge layer by:

$$C = \frac{\varepsilon \varepsilon^o A}{d} \qquad (4.9)$$

It should be noted that, despite the common use of the Helmholtz model for supercapacitors, often in these systems the capacitance is not constant with potential, as is assumed with the parallel-plate or Helmholtz model. This is particularly true of pseudocapacitive reactions where the capacitance can vary significantly with potential.

The energy stored and extracted from a double-layer supercapacitor is, as seen in Equation 4.4, related to the charge placed on the surface. The charge stored in a double layer is approximately 0.18 electrons per surface atom.[4] From this, it is obvious that for the same given surface area, the energy stored in the double layer is much less than that stored in a pseudocapacitive film that stores one or more electrons per surface atom. On the other hand, the extraction of stored energy (or charge) can be accomplished much faster in double-layer electrodes than pseudocapacitive films, as the formation and relaxation of the double layer occur in the order of 10^{-8} s, while the rate constant for faradaic reactions is typically in the range 10^{-2} to 10^{-4} s.[3] The higher rate of double-layer formation and relaxation means that the charge stored in the double layer can be added or removed more quickly than the charge stored in pseudocapacitive reactions, leading to higher power capabilities for double-layer supercapacitors (assuming that there are no significant pore effects, as described in Section 4.3.2).

4.3.1   The Classical Equivalent Circuit

The classic equivalent circuit used to model supercapacitors is a modified Randles circuit, where the solution resistance of the electrolyte is shown as a resistor (R_s) and the interface is modeled as a resistor (R_{ct}) in parallel to a capacitor (C), as in Figure 4.1a. The subscript "ct" in R_{ct} denotes that this is a charge transfer resistance, that is, this resistor models any noncapacitive faradaic reactions that occur across the surface, such as the reaction of an impurity or electrolyte. The capacitance (either double layer or pseudocapacitance) is modeled by the capacitor, as described above. This is the most basic model of a supercapacitor electrode, and truly only applies well for a planar double-layer capacitor. Systems with potential-dependent pseudocapacitance cannot be simply modeled using the Randles circuit because the capacitor in the Randles circuit implies a constant capacitance. For systems with pores, a more complex equivalent circuit (a transmission line) is used to accommodate the behavior of pores under applied AC voltage or current (see Section 4.3.2).

4.3.2   Modeling of Porous Electrodes

Highly porous electrodes are a source of extra complication for modeling of the double layer and developing equivalent circuit models for charging/discharging on a surface. This is because when a potential (or current) is applied to a porous electrode, it is not possible for the whole surface to respond instantly to this applied potential due to the distributed electrolyte resistance in the pores.[12,13] There have been many models proposed for porous electrodes. Typically, these models combine multiple RC circuits in various ways (in series and branched). The most common method used to model pores is using a transmission line, as proposed by de Levie.[12] In a seminal paper, de Levie showed that during charging or discharging of highly porous electrodes, the potential at the pore tip (closest to the other supercapacitor plate) responds more quickly than that at the surface deeper in the pores. Each unit length of the pore has some resistance associated with it, and deeper in the pores, the cumulative resistance increases. When an alternating potential is applied, the surface at the tip of the pore responds essentially instantly and follows the applied potential in both magnitude and frequency. However, further down the pores, the magnitude of the potential perturbation is not mirrored,[12] with lower magnitudes of AC potential at surfaces further down the pore; although, the frequency continues to mirror the applied frequency. Thus, during charging of a porous electrode, a distribution of potentials develops down the pores of the electrode.

de Levie modeled the equivalent circuit of a pore using a transmission line (Figure 4.1b), where each small pore section (dz, where z is the distance from the pore mouth) is modeled using a capacitor that models the capacitance of the surface and an in-series resistor to model the resistance of the solution in the pores.[12] Although the electrode material of which the pore walls are composed has some resistance, the resistivity of the electrode material is often much lower than the electrolyte resistance in the pores, and is therefore usually ignored.

In terms of the electrolyte resistance in the pore, the relationship of resistance is similar to that seen in a wire; both longer and narrower pores result in a higher incremental solution resistance, and therefore an increase in cumulative resistance. The presence of bottlenecks in the pores may also influence the resistance considerations/profile of a pore.[14]

Often, a porous material has micropores in the walls of the meso- and macropores, giving a fractal-type geometry of the surface. These more complicated pore structures are modeled using nested transmission line circuits, where a pore is placed parallel to the capacitor for the wall, such as the example discussed by Itagaki et al.[15]

4.4 Pore and Film Resistance Effects on Key Descriptors and the Impact of Rate

The rate of charging and discharging of a supercapacitor plays an important role in many of the descriptors described in Section 4.2. The rate effects are particularly evident in porous electrodes or pseudocapacitive films with resistance to ion migration. The effects of rate are briefly outlined here.

4.4.1   Resistance and Voltage

As is obvious from the above description of the pore effect, pores have higher resistances due to the cumulative solution resistance. While the pore resistance itself is not influenced by high rates (i.e., higher currents), the voltage drop, or iR-drop, is more pronounced at higher rates. In terms of porosity, the pore resistance relationship is similar to that seen in the resistance of a wire; longer pores result in higher cumulative resistance, as do narrower pores. Thus, longer or narrower pores will result in more voltage drop down the pore length. The presence of bottlenecks in the pores may also dominate the resistance considerations/profile of a pore.[14]

Similarly, with pseudocapacitive films, the ion migration limitations (resistance) can be significant for thick films. This film resistance leads to the surface of the film being reacted more readily during charging or discharging, resulting in a different oxidation state for the surface species (e.g., different Ru or Mn oxidation states in $RuO_2.xH_2O$ or $MnO_2.xH_2O$) films than in the bulk.[16,17] Indeed, for films (like manganese oxide) where the film resistance is relatively high, only a very thin layer of the film is used during charging/discharging.[17] Often, thin films are used to limit this resistance.[8]

The resistance effects seen in porous electrodes and pseudocapacitive films limit the rate at which the ion may migrate into a pore or through the film, and thus may limit the rate of charge transfer in a supercapacitor. In turn, higher rates of charge or discharge results in the material being used less effectively than at lower rates.

4.4.2   Charge and Capacity

For planar double-layer electrodes, there are no rate considerations.[1] Increasing the porosity of a material increases the surface area and therefore increases the double-layer charge that can be stored (capacity) on an electrode in a given voltage window. Similarly, to a first approximation, increasing the amount of film in pseudocapacitive materials increases the amount of charge that can be stored (i.e., increases capacity). For porous materials, increasing the surface area, without a concomitant increase in volume, is often accomplished by forming small pores in the material. However, for porous electrodes and those based on pseudocapacitance, the full capacity may not be realized at high rates, as the full surface area (or volume) of the electrode may not be charged because of resistive effects down pores or through films.[6,16,18] At higher rates of charge/discharge, less of the pore or pseudocapacitive film reaches the desired potential, and the "penetration depth" decreases. This leads to a lower surface area being charged, and therefore a lower charge or capacity at higher rates.[1] At the highest limit, only the outer portion (or face) of the electrode will be charged and this low surface area results in a small charge or capacity. At lower rates, more of the electrode depth is charged, resulting in a higher capacity or charge.

As the resistance in a pore is influenced by the electrolyte concentration, lower electrolyte concentrations may lead to decreased penetration depths, and therefore may lead to lower capacities at high rates, for similar reasons described above.[1] Similarly, using a lower temperature results in a higher solution or film resistance, and therefore a lower "penetration depth" for the voltage, and a lower charge or capacity.

4.4.3 Capacitance

The influence of the solution resistance in the pores of porous electrodes or the film resistance in pseudocapacitive electrodes can influence the capacitance measured for an electrode at different rates. Low-resistance films, or wide pores, allow for the removal of higher amounts of charge at a particular rate. Generally, for high-resistance films, materials containing small pores, or at high rates, the penetration depth into the pores or the depth of reaction into the film drops, resulting in lower effective electrochemical surface areas, and therefore lower capacitances.

4.4.4 Coulombic Efficiency

While both capacity and capacitance of a system depend on the rate and resistance of the cell, the CE typically does not have any rate effects; assuming the system is charged and discharged at the same rate, the same charge will be placed on the surface and removed from the surface. Even though the penetration depth of a porous electrode or pseudocapacitive film may change with rate, if the same rate is used, then the penetration depth is the same for both charging and discharging.

4.4.5 Time Constant

The charging/discharging rate, porosity, and film resistance all have a significant impact on the time constant (RC) for a supercapacitor since different penetration depths into a pore or film result in different cumulative resistance and capacitance. Indeed, for porous electrodes and pseudocapacitive films there is often no one time constant but rather a range of time constants. Nevertheless, a characteristic time constant for the system is often determined (typically using EIS, see Section 4.10.3) to allow for comparisons between electrode materials or supercapacitor systems.

4.4.6 Energy

The resistance in the system has a significant impact on the energy lost through dissipation in the cell (see Section 4.2). A low value for resistance (pore solution resistance or ion-migration resistance through a pseudocapacitive film) allows for easier ion transport into the material (i.e., a higher penetration depth) and therefore a larger degree of charge storage and higher energy. This is particularly true of porous electrodes, where the resistance down the pore is a cumulative resistance. Upon application of a voltage, the current decreases down the pore while the cumulative resistance increases. Therefore, there is no single iR value for a porous system,[2] but rather a range of ohmic loss values varying down the pore. The energy lost through the pore resistance is then dependent on the penetration depth of the voltage into the pore.[2]

4.4.7   Power

Highly porous electrodes cannot respond as rapidly to the application of current or voltage as planar electrodes, as explained earlier. Practically, this corresponds to a lower maximum current at which the full surface may be charged, and since power depends on current, it results in a lower power. For situations where high-power capabilities are desired, large (wide, low resistance) pores are used, often formed by templating methods or the incorporation of carbon nanotubes in the material.[19–22] Small pores may be used for high power, so long as the pore is relatively short (minimizing cumulative resistance). Complex architectures, combining large pores for maximum current flow with short, small pores to provide a high–surface area, are used to provide higher energy without sacrificing much in terms of power.

4.5   Two-Electrode versus Three-Electrode Measurements

Supercapacitor measurements are conducted in two different ways: two-electrode and three-electrode measurements. In two-electrode measurements, also termed "full-cell" measurements, each electrode in the measurement corresponds to one of the electrodes used in a real supercapacitor. The electrode potentials are allowed to move as they would in a real supercapacitor; that is, during charging, the potential of the two electrodes move away from each other (the cell voltage increases); during discharge, they move toward each other (the cell voltage drops). Since the potential of both electrodes changes throughout the charging/discharging process, the potential of each electrode is not known exactly, and only the cell voltage is known. Two-electrode measurements provide a more realistic picture of the characteristics of the supercapacitor system; however, they do not allow for a convenient method to identify the redox reactions occurring on each electrode during a charge/discharge process.

   In a three-electrode measurement, only one supercapacitor electrode is examined, namely, the working electrode. A counter electrode is present that allows for the free passage of current, but does not otherwise influence the electrochemistry. The third electrode in a three-electrode system is the reference electrode. The potential of this electrode is fixed at a known value and does not deviate during the experiment. The potential of the working electrode is measured with respect to this reference electrode and as such the potential of the working electrode is known exactly. In this way, the capacitance of each supercapacitor electrode may be determined, as well as the identity and potential of any redox reactions occurring on the electrode surface.

4.6 Asymmetric versus Symmetric Capacitors

In symmetric supercapacitors the two electrodes are made of the same material, whereas in asymmetric supercapacitors the two electrodes are composed of different materials (e.g., one electrode may be a metal oxide or conducting polymer pseudocapacitive electrode, while the other electrode is a carbon double-layer capacitive electrode). The characterization techniques described in this chapter are often used to examine asymmetric systems because this gives a more realistic description of the characteristics of a true asymmetric supercapacitor. Nevertheless, it should be noted that, just like the two-electrode case described in the previous section, in asymmetric systems the supercapacitor response is influenced by both the positive and negative electrodes at any particular voltage, and thus, the specific characteristics of each electrode are lost in the measurement of the asymmetric system. Changes in the response of the system are then difficult to assign to changes in one or the other of the electrodes. Nevertheless, the response of the system will be more representative of the characteristics of the cell as a whole; one can get a better idea of the potential that will result in the highest charge storage, the practical and stable voltage window, etc. The issue of not being able to assign changes in the response to a specific electrode can be alleviated by using a third electrode (as the reference electrode) in the system and tracking the response of each electrode separately. In this way, it is possible to examine the important characteristics of each electrode but in a realistic system, with all of the limitations and changes that would be seen in a real supercapacitor.

4.7 Cyclic Voltammetry

In cyclic voltammetry the potential of an electrode (or voltage of a system) is changed linearly between two potential extremes and then returned to the initial potential to complete the cycle. The cycle is often repeated a number of times. The current that passes during this potential cycle is recorded as a function of the potential. The rate at which the potential is changed is called the sweep rate (v). Cyclic Voltammograms (CVs) are the resulting current versus potential plots. There is some variation on how CVs are plotted, but the International Union of Pure and Applied Chemistry (IUPAC) convention states that the oxidation current is considered to be a positive current, with the reduction current being negative, and that increasing positive potentials be plotted toward the right of the graph.

Often charging and discharging of a battery or supercapacitor is given in terms of "C-rate." The relationship between C-rate and sweep rate from CV is as follows:

$$\text{C-rate} = 3600 \times \frac{v}{V} \qquad (4.10)$$

where V is the maximum voltage of the system. Even relatively slow sweep rates for cyclic voltammetry are equivalent to relatively high C-rates. For instance, for an aqueous supercapacitor with a 1-V voltage window, a sweep rate of 1 mV/s results in a C-rate of 3.6.

4.7.1 Requirements for a Material to Be Considered Capacitive by Cyclic Voltammetry

Cyclic voltammetry is often used to evaluate whether an electrode material is capacitive. In order for a material to be considered a suitable supercapacitor material, it should have electrochemical properties similar to the electrical properties of a typical capacitor. The CV of a typical capacitor is a rectangle, as the capacitance is independent of potential; the CV is a mirror image about the zero current line and has vertical current switches at the two potential extremes (i.e., when the potential sweep direction is changed, the current changes sign but not magnitude). Additionally, when the capacitor is charged at different rates (different sweep rates), the CV current is directly related to the sweep rate. These three criteria (mirror image around zero-current line, vertical current switches at potential extremes, and linear relationship between current and sweep rate) are the main requirements to determine whether something may be called "capacitive." In practice, the criteria of the mirror image and vertical current switches may be relaxed somewhat, though, as discussed in more detail in Sections 4.2.12 and 4.7.6, variations from these criteria result in materials with lower energy or power densities.

Double-layer capacitive electrodes, such as carbon materials, typically demonstrate all three capacitance criteria, as typified by Spectracarb™ 2225 (Engineered Fibers Technology, Shelton, CT, USA) in 1 M NaOH (Figure 4.2).[23] Pseudocapacitive materials must also satisfy these three criteria, although

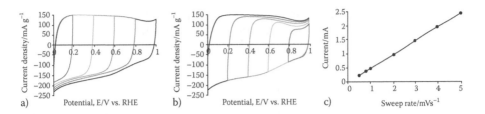

FIGURE 4.2
Spectracarb 2225 carbon cloth electrode in 1 M aq. NaOH electrolyte showing the rectangular CV shape with vertical current switches at potential extremes a) and b) and the linear dependent of current on sweep rate c). (From Andreas H.A. and Conway B.E., *Electrochim. Acta*, 51, 6510–6520, 2006. With permission.)

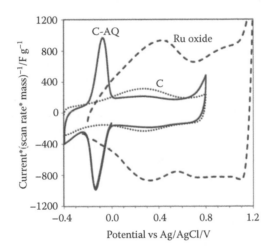

FIGURE 4.3
CVs of ruthenium oxide and anthraquinone-modified Spectracarb 2225 carbon (C-AQ) in 1 M aq. H_2SO_4, showing the reversible pseudocapacitive CV shape. (From Algharaibeh, Z., et al., *J. Power Sources*, 187, 640–643, 2009. With permission.)

these materials may have CVs that do not appear rectangular if there is a capacitance dependence on potential. For example, hydrous ruthenium oxide is a well-known metal-oxide pseudocapacitive material that shows the three characteristics outlined earlier and has a fairly rectangular CV shape, although some structure is seen, as shown in Figure 4.3.[24] Anthraquinone-modified carbon also exhibits a pseudocapacitive CV, as seen in Figure 4.3; however, the CV is not rectangular as there is a strong capacitance dependence on the potential (i.e., when the anthraquinone undergoes its oxidation and reduction reactions at −0.11 V, there is an increase in the capacitance). Both the RuO_2 and anthraquinone-modified carbon CVs are mirror images about the zero-current line and have vertical current lines at the potential extremes, confirming these are pseudocapacitive materials. Similar pseudocapacitive

CVs have been observed with other metal oxides, such as manganese oxide, and conducting polymers, such as polyaniline and polypyrrole, as described in more detail elsewhere in this book.

4.7.2   Stable Voltage Window

Cyclic voltammetry is the most common method used for determining the stable voltage window of a supercapacitor material. Typically, the CV of the material is examined over the possible range of potentials that the material may encounter when used in a supercapacitor. Additionally, the stable potential window is examined over extended times for changes in CV shape and size. The analysis of these shape and size changes will depend on the system under study, although a few examples are provided here. A change in size may indicate that the electrode material may be dissolving, or losing electrical contact with the surface, such as that seen in polymers upon repeated cycling, due to volume changes of the polymer during cycling.[25] A change in the CV shape, particularly the resistance evident in the CV (as described in Section 4.8.10), may indicate that the material is breaking down; for example, the breaking of the conductive graphene sheets in carbon due to oxidation.[1] Alternately, an increased resistance may indicate gas formation during overcharging, which may block the pores of a porous material. The voltage window may also be limited by dissolution of electrode material,[7] decomposition of aqueous impurities in an organic electrolyte,[26,27] or electrolyte decomposition in acetonitrile.[28]

4.7.3   Evaluating Capacitance by CV

Capacitance is defined as the amount of charge stored at a particular voltage. Often, authors will use CVs to report just one value for capacitance. Sometimes this value is an average capacitance, but more often it is the maximum capacitance. However, CVs can be used to give a more useful representation of capacitance: the differential capacitance (C_{diff}). The differential capacitance provides a measurement of the capacitance at each particular voltage and can be determined from CV by dividing the measured current (I) by the sweep rate (v):

$$C_{diff} = \frac{dI}{dv} \tag{4.11}$$

When differential capacitance is calculated from Equation 4.11, cathodic currents result in a negative differential capacitance. While a negative capacitance has no physical meaning, the negative sign of the differential capacitance calculated in this manner denotes simply that it is the capacitance when a negative (or reduction) current is flowing into the working electrode of a three-electrode system.

The differential capacitance is useful for supercapacitor materials because the capacitance is often not constant with voltage, as is expected with typical capacitors, but may vary greatly. This is particularly true of pseudocapacitive electrode materials such as metal oxides and conducting polymers.

For ideal supercapacitors, there is no influence of the CV sweep rate on the capacitance of the system and, in fact, is one of the requirements of a capacitive material. The capacitance independence of rate is particularly true of planar double-layer capacitors, where the double layer forms and relaxes very quickly (10^{-8} s),[3] and therefore the electrode is able to respond almost instantaneously as the potential is changed during the potential sweep (note that the pores in high–surface area double-layer capacitors cause the capacitance to depend on sweep rate, as described in the next paragraph). Faradaic reactions are slower than double-layer formation, often requiring 10^{-2} to 10^{-4} s,[3] and therefore at very high sweep rates capacitive faradaic reactions may become rate limited.[4]

In both double-layer and pseudocapacitive systems, care is put into choosing conductive electrolytes, electrode materials, separators, and current collectors to minimize the resistance in the cell which would otherwise result in a drop in voltage (iR-drop) in the cell. This iR-drop results in less actual voltage across the electrodes in the cell for the same applied voltage to the system. A drop in actual voltage applied to the supercapacitor electrode results in less charge on the electrodes than if the whole applied voltage was applied to the electrodes. Since it is the applied voltage which is used in calculating the system capacitance, the capacitance may appear to be less in a system with a large iR-drop than for a system will little resistance. The iR-drop in pores must also be considered, and increases in solution resistance lead to smaller penetration depths, lower electroactive surface area, and therefore lower capacitance (see Section 4.4 for a discussion on pore effects). Higher sweep rates result in higher currents, and therefore greater iR-drops, leading to decreased capacitances.

For a CV recorded for a system which has both significant double-layer and pseudocapacitive components, the capacitance of each component can be determined from the CV only if there is a portion of the CV which does not have significant pseudocapacitive contribution (i.e., there is a portion of the CV with the appearance of the small, rectangular double-layer charging profile). The differential capacitance of each portion of the CV can be calculated. It must then be recognized that the capacitance calculated for the region exhibiting pseudocapacitance is actually composed of capacitance for both the pseudocapacitive reaction and the double layer, and therefore the double-layer capacitance can then be subtracted from the combined capacitance to determine the capacitance due to pseudocapacitance alone. In the pseudocapacitive region, the assumption must be made that the double-layer capacitance has the same value within this region, as it does in the double-layer-only portion of the CV. As the pseudocapacitive capacitance is typically much larger than the double-layer capacitance, this assumption can often be safely made, as the small variations in the double-layer capacitance do not greatly influence the values calculated for pseudocapacitance. Note that for systems where there is no portion of the CV that exhibits solely double-layer charging, the method just described cannot be used to separate the capacitance due to pseudocapacitance from that of double-layer charging. In these cases, the capacitance from double-layer charging may be estimated (often using the value of 0.18 electron per surface atom), but the error associated with this estimation and the calculated pseudocapacitance value is large.

4.7.4   Capacity/Charge

The charge (Q) can be calculated from a CV through integration under the curve followed by dividing by the sweep rate (v). Alternately, if the CV has already been converted to a differential capacitance (C_{diff}) profile (using Equation 4.11, wherein the current was divided by the sweep rate) then integration under the curve results directly in the charge (or capacity):

$$Q = \int I dV * \frac{1}{v} = \int C_{diff} dV \qquad (4.12)$$

At higher sweep rates and higher resistances, the CV develops a tilt, which indicates that resistance is influencing the electrochemistry. The capacitance calculated from the CV also becomes smaller since the pores become less active at high rates; there is a lower penetration depth, resulting in a smaller electroactive area and consequently a lower capacitance.

For CVs that are recorded for materials which have significant contributions from both pseudocapacitive and double-layer charging, the charge of each process can be calculated from the CV only if there is a portion of the CV that exhibits solely double-layer charging. The current in the double-layer region is then assigned as the double-layer charging current throughout the CV, and the double-layer charge is calculated using this current and the whole potential window. This double-layer current is also used as the background current for the calculation of the pseudocapacitive charge, where the double-layer current is subtracted from the current in the region of pseudocapacitance prior to integration through this region. Once the background is subtracted, the remaining charge is assigned as the charge associated with pseudocapacitance.

For processes that lead to loss of capacitance, the CV often exhibits a decrease in current with cycling as portions of the electrode lose activity. This can be the case for systems, such as electroactive polymers, which because of their volume changes can become electrically disconnected over multiple cycles, and the CV current at any given potential drops as more of the film becomes disconnected.

4.7.5 Coulombic Efficiency

CE is calculated using the charges from the CV as described in Section 4.7.4 and Equation 4.3. The CE of a material should not be influenced by the sweep rate or changes in the electrolyte resistance. The exception may be if there is an irreversible parasitic faradaic reaction occurring in parallel to the capacitance. In this situation, the CE will be sweep rate dependent, with higher sweep rates resulting in CEs closer to 100% than those seen at lower sweep rates. At high sweep rates, the irreversible reaction may be unable to react sufficiently fast (i.e., essentially "outrunning" the irreversible faradaic reaction), resulting in a lower contribution to the charge from this irreversible reaction. At lower sweep rates, there will be a larger contribution from the irreversible reaction and the CE will deviate from 100%.[29] A link may be made between poor CEs and shorter system lifetimes, because as these irreversible reactions are sometimes linked to the eventual failure of the energy storage system[2] (such as electrolyte decomposition or active material decomposition/dissolution[7]). Smith and co-workers are examining methods to use precisely measured CEs to predict life cycle in Li-ion batteries.[30] This may also be applied to other energy storage systems in the future.

4.7.6 Energy and EE

Energy is related to the potential at which the charge is added or removed from the surface (Equation 4.4). To calculate the energy from the CV, one may plot the charge as a function of potential (charge is calculated by multiplying the current by time of charging) and integrate under the curve. EE is then the percentage ratio of the energy delivered on discharge to the energy required during charge.

Energy and EE are both strongly dependent on rate since ohmic, activation, and concentration polarization losses all contribute to dissipative energy losses in the system (as described in Section 4.2). Even electrochemically reversible reactions have a limit at which the sweep rate begins to be too similar to their rate constant and the peaks in the CV shift apart.[4] The system becomes irreversible under these conditions and therefore the energy decreases.

4.7.7 Resistance

An increased resistance in a system is most evident in the shape of the CV at the currents of the potential extremes. At higher resistances, the current response is sluggish and this results in some curvature in the CV at the potential limits. For porous and ion-migration limited systems, this curvature of the CV indicates the charging time for the system.[31] In extreme cases, the CV becomes more sloped and begins to resemble the straight sloped line CV of a resistor.

Reduced electrolyte conductivity and lower temperatures are expected to result in more resistance in the cell and more curvature in the CV.[1] Higher porosity leads to a higher surface area, and therefore higher capacitance and capacity, but the distributed resistance in the pores leads to a more resistive CV, or a lower sweep rate, at which the resistance effects are seen. Factors that increase the resistance in the pores, such as narrower or longer pores, or the presence of bottlenecks, will result in similar CV distortions.

Some irreversible processes can lead to an increase in the resistance of the electrode material (e.g., the overoxidation of a carbon electrode), and this presents as an increasing tilt in the CV and a delay in current recovery at the potential extremes in the CV (i.e., more curvature at the potential extremes).

Although resistance can be seen in a CV, typically the CV is not used to calculate a value for the resistance; instead, galvanostatic charging/discharging experiments (Section 4.8.7) or EIS (Section 4.10) are used to determine the system resistance.

4.7.8   Power

Power is typically not calculated from a CV, but can be done by integrating under the curve. The more typical, and simple, method to determine the power of a supercapacitor system is to study the system using galvanostatic charging/discharging (Section 4.8.6).

4.8   Galvanostatic Charge/Discharge

In galvanostatic (or constant current) charge/discharge experiments, a known, set current (I) is applied to an electrode and the potential is recorded over time. When the electrode reaches a desired potential, the current direction is changed (the same current but opposite in sign is now applied) and again the potential is recorded. Often, the potential will be cycled between these two potential limits. Typically, strict potential limits are placed on the experiment to avoid overcharging of the electrode or system, which can result in electrolyte or electrode decomposition. The galvanostatic charge/discharge curves plot the recorded potential as a function of time (t) (Figure 4.4) or as a function of charge (Q), using $Q = It$ (where I is constant).[1] Similar experiments can be conducted where only the discharge curve is recorded. In this case, the supercapacitor has been previously charged to the desired voltage and the current is stopped (current = 0 A). Then the supercapacitor is discharged and the voltage is recorded during the discharge. These curves can be analyzed in the same way as the

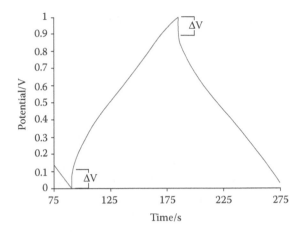

FIGURE 4.4
Galvanostatic charge/discharge curve for Spectracarb 2225 in 0.5 M aq. H_2SO_4 using a 6-mA charge/discharge current. δV shows the instantaneous drop in voltage when the current direction is changed. (From Andreas H.A. and Conway B.E., *Electrochim. Acta.* 51, 6510–6520, 2006. With permission.)

charge/discharge experiments, except for one small consideration when using these curves to determine the equivalent series resistance (described in Section 4.8.10).

4.8.1 Requirements for a Material to Be Considered Capacitive by Galvanostatic Charge/Discharge

For a material to be considered capacitive using galvanostatic charge/ discharge curves, the curves should be symmetric between charging and discharging, preferably with linear slopes to indicate constant capacitance as a function of potential. The linear slope requirement may be relaxed, but in order to achieve high coulombic and energy efficiencies, the mirror-image charge and discharge curves must be maintained. Ideally, the potential drop at the current switching point is very small or absent, as a large instantaneous potential drop at these points indicates significant equivalent series resistance in the cell, as discussed in Section 4.8.10.

4.8.2 Appearance of Curves When Capacitance Is Constant with Potential

For systems where the capacitance is constant over a range of potentials, the galvanostatic charge/discharge curves are triangular with linear sides (Figure 4.4). As described in Section 4.8.5, the slopes of these linear sides are equal to the inverse of the capacitance when the x-axis is plotted as charge. These linear triangular shapes are often seen with double-layer capacitors (where the double-layer capacitance is constant, or almost constant, at various potentials) and some pseudocapacitive metal oxides, such as $RuO_2 \cdot xH_2O$, which exhibits an almost constant capacitance with voltage.

4.8.3   Appearance of Potential-Dependent Pseudocapacitance

In pseudocapacitive systems, where capacitance varies with potential, a change in slope corresponding to the capacitance change is evident in the galvanostatic charge/discharge curve. These nonlinear galvanostatic charge/discharge curves are common with pseudocapacitive materials, such as metal oxides and electroactive polymers, where the capacitance derived from an electrochemically reversible faradaic reaction can be strongly potential dependent. The tangent at each point on the curve can be used to determine the capacitance at each particular potential (see Section 4.8.5). Thus, curvature in galvanostatic charge/discharge curves is often indicative of a pseudocapacitive component if the capacitance change is mirrored on both the charge and discharge curves. Curvature in the profiles can also happen with parasitic redox reactions, which are not pseudocapacitive, and since these reactions are not electrochemically reversible, the capacitance changes are not the same during charge and discharge and this is reflected in a lack of mirror image between the charge and discharge curves. Galvanostatic charge/discharge curves that are not mirror images between charge and discharge result in lower coulombic and energy efficiencies for the system.

4.8.4   Appearance of Porous Effects

The distributed resistance of a porous electrode or pseudocapacitive film also may cause curvature to the voltage versus time profile of galvanostatic charge/discharge curves.[2] A porous electrode was modeled using a transmission line hardware circuit consisting of five RC elements by Pell et al.[2] During galvanostatic charging and discharging, the current is split among the five capacitors, and since the capacitors further down the circuit begin to charge later in time during the charging cycle, the potential of the top capacitor (modeling the pore mouth, where the potential is measured in a real system) increases, but not linearly.[2] As the capacitance in the system increases (more capacitors start being charged), the profile exhibits a drop in slope, as is consistent with the expected inverse relationship between capacitance and slope. Essentially, the current is split between the capacitors, but the current moving into each capacitor (or portion of the pore) changes with time, and therefore, the voltage profile is not linear with time.[2]

Porous electrodes also, initially, result in nonmirror-image charge/discharge profile.[2] During the first few galvanostatic charge and discharge cycles, the inner portions of the pore are being charged, but are unable to donate charge back to the surface during discharging. After several cycles, the profiles reach a steady state (each subsequent profile is the same as the previous profile) and become symmetric. The number of charge/discharge cycles required to reach this steady state is rate dependent with slower rates (lower applied constant currents) resulting in the steady state being reached more quickly.[2] Pell also showed that the energy losses associated with porous electrodes decreased with cycling, and therefore, galavanostatic charging/ discharging should be conducted for a number of cycles in order to examine how the profile changes with cycling.

4.8.5 Capacitance

To determine the capacitance from the voltage versus time curve (Figure 4.4), the inverse of the slope is multiplied by the applied current (recall Equation 4.1). If the curve is plotted with voltage as a function of charge, then the inverse of the slope gives the capacitance directly. The average capacitance is typically taken from the linear portion of the curve, and the differential capacitance can be determined using the tangent at each potential. Because the capacitance is inversely related to the slope of these curves, an increase in capacitance, such as would be seen through a pseudocapacitive reaction, leads to a drop in the slope of the plot, often seen as a flattening or curvature in the plot. Similarly, with porous electrodes, as the charge redistribution results in charge moving further down the pore (increasing the electroactive surface area and therefore the capacitance) there is a flattening of the galvanostatic profile.[2]

4.8.6 Capacity/Charge Density

The charge or capacity of the supercapacitor can be determined from the galvanostatic charge/discharge plot by multiplying the constant current applied by the time required to complete the charging or discharging to arrive at the charging charge and discharge charge, respectively.

As discussed previously, a drop in electrochemically active surface area caused by an increase in the resistance of the electrolyte in the pores or in the capacitive film results in lower amounts of charge storage, and lower charge densities or capacities. On galvanostatic charge/discharge curves, this appears as a decrease in charge or discharge time.

Irreversible processes lead to a loss of capacity present in galvanostatic charge/discharge curves and a drop in the time required to fully charge or discharge the surface. As less charge can be accommodated on the electrode, less time is required at the applied current to fully charge/discharge the system. Losses in capacity such as this can be attributed to some irreversible process, which make portions of the material inactive, such as the electrical disconnection of polymers during charging/discharging due to volume changes.

4.8.7   Coulombic Efficiency

The charges calculated from the galvanostatic charge/discharge curves may be used to calculate the CE. Because the current applied is constant throughout the experiment, the voltage–time curves can be easily used to see CEs; for a CE of 100%, the charge time is equal to the discharge time. If the charge during either charging or discharging is larger, it results in a non–mirror-image curve and differences in the charge and discharge time, and thus poor CEs are easily seen with these curves.

Similar to what was seen with cyclic voltammetry, CE should be independent of the rate (or current) of charge and discharge. CE will, however, display rate dependence in the presence of an irreversible faradaic reaction (as described in Section 4.7.5).

4.8.8   Energy and EE

The energy of the system is one of the characteristics most commonly determined from galvanostatic charge/discharge curves, particularly for use in Ragone plots (described in Section 4.9). The energy can be calculated from the voltage–time curve by integrating under the charge or discharge portion of the curve and multiplying by the applied current. Alternately, if voltage is plotted versus charge, energy can be determined directly by integrating under the curve. As described in Section 4.2.11, 75% of the energy of a supercapacitor system is delivered in the first half of the voltage window, and this can be easily seen from the galvanostatic voltage–time curves (Figure 4.4), where 75% of the area under the discharge curve is a result of the first half of discharge. For constant-current experiments, the practical considerations are that the time required to deliver 75% of the energy is equal to the time required to deliver the last 25% of the energy, highlighting why supercapacitors are practically discharged to only half of their initial voltage.[1]

The energy loss experienced in porous electrodes is highly rate dependent. At low rates, more of the inner pore is charged, and therefore, the energy losses from the inner pore contribute more significantly to the total energy losses than at high rates.[2] However, overall the energy is much larger (due to the higher surface area associated with the inner pore being charged), and therefore, even with the greater energy loss, the resulting energy is larger at the low rates. Similar reasoning can be applied to energy considerations for pseudocapacitive materials with ion-migration resistance in the film.

EE is calculated as described above in Section 4.2.12. Galvanostatic charge/discharge curves, which are not mirror images, may indicate a low EE for similar reasoning as described in Section 4.2.12, where irreversibility in the system results in charge being placed on the supercapacitor at a high voltage but removed at a low voltage leading to a higher charging energy than the discharging energy. A change in the slope (capacitance) of the charge curve must be mirrored at the same potential in the discharge curve (i.e., charge must be removed from the capacitor at approximately the same potential as it was placed on the capacitor) in order for the EE to be 100%. Thus, a low EE is easily seen in the curves resulting from galvanostatic charge/discharge measurements.

4.8.9 Power

Power is another characteristic typically determined from galvanostatic charge/discharge curves and can be determined as a function of SOC by multiplying the measured potential and applied current. The power calculated by this method is typically used in the construction of Ragone plots (described in Section 4.9). As with energy, the power losses experienced in porous electrodes and resistive pseudocapacitive films are highly rate dependent. Since power loss is related to I^2R (Equation 4.8), higher currents (rates) result in higher power losses.

4.8.10 Resistance

The equivalent series resistance of a cell can be easily calculated from constant current charge/discharge curves. At the point where the current direction changes, there is an instantaneous drop in the potential (see ΔV in Figure 4.4), which is equal to $2IR$. Since the current is known (applied by the researcher), the resistance can be calculated. Note, however, that if the experiment is composed of only the discharging portion (i.e., the capacitor was charged and then the current was set to zero, even briefly) then the potential drop is equal to IR.[32] In the charge/discharge experiment, there is a voltage drop through the equivalent series resistance as the charge moves in one direction and an equal drop occurs when the charge direction is changed, which leads to the $2IR$ voltage drop. When only the discharge curve is recorded, the energy lost is only that for the movement of ions in one direction, and therefore the potential drop is equal to IR.

Processes such as gas evolution (blocking pores) and disruption of the electrical connectivity of the electrode material (e.g., overoxidation of graphene sheets in carbon, delamination or dissolution of electrode material, disconnection of material due to repeated volume changes in electroactive polymers, etc.) can increase the resistance of the system. These increases in resistance are easily seen in galvanostatic charge/discharge results as an increase in the initial potential drop in the charge/discharge curves.

4.9 Ragone Plots

There are three main types of Ragone plots, though all are plotted with the log of power on one axis and the log of energy on the other. Typically, Ragone plots are constructed from galvanostatic charge/discharge profiles (see Section 4.8), although they are sometimes constructed from Electrochemical Impedance Spectroscopic data. The three types of Ragone plots typically seen in the literature are those that plot: (1) the ranges of energy and power densities available for different types of energy storage systems; (2) the energy and power densities for a given system at various states of charge; and (3) the maximum energy and power densities for a given system charged at various constant currents. The first type of Ragone plot, which compares systems, is beyond the scope of this chapter. Types 2 and 3 are covered in subsequent sections.

4.9.1   Ragone Plots as a Function of SOC of a System

As described earlier, one of the inherent characteristics of supercapacitors is that the potential (or voltage) changes with the SOC. Since both energy and power are related to potential, the energy and power densities plotted in Ragone plots are expected to change as a function potential, and therefore as a function of the SOC. Indeed, as seen in Figure 4.5, both power and energy increase almost linearly with increasing SOC.[1]

Commonly, these curves have a hooked end at the highest power since at high rates the charge will be removed from the outer portion of the electrode first (where there is the least solution resistance); however, the outer portion of the electrode stores proportionally much less energy than the inner portion (the pores of the electrode or the bulk of the pseudocapacitive film), and therefore, only small amounts of energy are available at these high powers.

As expected from the discussion in Section 4.4, the maximum energy increases with decreased resistance, because of a larger penetration depth into the electrode material. As the resistance is increased, or at higher rates, the maximum energy falls since not all of the surface is being accessed by the penetration depth. At very high currents or high resistances, only very small energy is seen since only the outer portion of the film is used,[1,13] and this is a much smaller surface area than is available when the whole film or porous electrode area is used. At very low resistances, when the whole surface is accessed at a given rate, the maximum energy becomes independent of rate.

Power increases at increasing rates, as expected from Equation 4.8. The influence of current on the power is more evident at low resistances, where there is only a small associated drop in potential[1]. At higher resistances, where the *IR* drop becomes significant, the increase in power with increasing current is less obvious.[1]

4.9.2   Ragone Plots at Full Charge for Different Constant Currents

Often, Ragone plots are shown based on the power and energy of a system at a SOC of 100% (full charge). Ragone plots of this type for supercapacitors typically have a hooked shape, such as that seen in Figure 4.6. The hooked

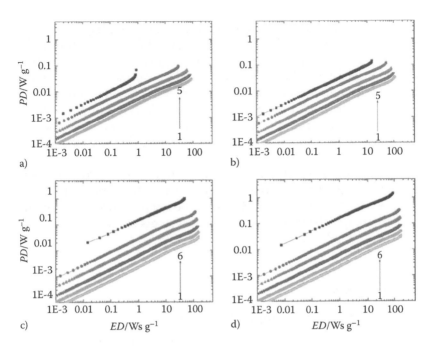

FIGURE 4.5
Ragone plots as a function of SOC for Spectracarb 2225 in H_2SO_4 at concentrations of a) 0.01 M, b) 0.05 M, c) 0.5 M, and d) 5 M. The data were derived from galvanostatic charge/discharge experiments using current densities of (1) 1 mA, (2) 1.5 mA, (3) 2.5 mA, (4) 5 mA, (5) 10 mA, and (6) 50 mA. (From Niu, J. et al., *J. Power Sources*, 135, 332–343, 2004. With permission.)

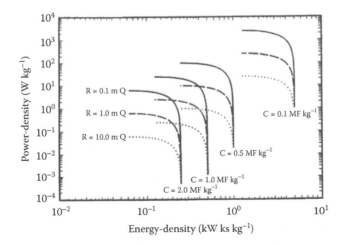

FIGURE 4.6
Ragone plots for fully charged 1 MC/kg capacitors (with nominal capacitances of 0.1 MF/kg, 0.5 MF/kg, 1.0 MF/kg, and 2.0 MF/kg) discharged through various resistances. (From Niu, J. et al., *J. Power Sources*, 156, 725–740, 2006. With permission.)

shape is due to the resistive effects in these systems (such as the pore effect and resistance to ion migration through pseudocapacitive films). As the resistance in a system is reduced, by using higher electrolyte concentration or higher temperatures, both the energy and power increase.[1] At high rates, which result in high powers, the penetration depth into the porous electrode or pseudocapacitive film drops, resulting in a lower charge (as described above), and therefore a lower energy. At low rates, the whole surface or film can be charged, resulting in a high energy, but the low current results in a lower power.[33]

While these Ragone plots show information for only a 100% SOC system, the trends between fully charged systems seen in these plots are preserved at lower SOCs;[1] therefore, these plots can be usefully used to compare the power and energy characteristics between systems or electrodes.

4.10   Electrochemical Impedance Spectroscopy

EIS is used to determine quantitative information on the different resistive and capacitive components of a system as well as their frequency dependence—assuming, of course, that an appropriate and real-life relevant equivalent circuit is used for the calculation of these values (see Equations 4.13–4.15). EIS is often used to quantify the equivalent series resistance, determine the charge transfer resistance and system time constant, elucidate pseudocapacitive behavior, and examine the role of porous electrodes.[4] EIS experiments are typically not used to provide any information on the charge stored on a supercapacitor electrode, the CE, energy, or EE. For those characteristics, galvanostatic charge/discharge or CV is used.

In EIS experiments, a small sinusoidal potential perturbation of 5 to 10 mV is applied to the system and the resulting current is recorded. Alternately, a small alternating current may be applied and the resulting AC voltage is recorded.[34] The frequency of the applied alternating signal is varied with frequencies between 10 mHz and 1 MHz being common (though, usually not the whole range of these frequencies is used). Sometimes a superimposed DC potential or current is also applied to the system. The small magnitude of potential or current perturbations is used to ensure that the current is linearly proportional to the potential. EIS experiments may be run in either a three-electrode configuration or two-electrode, full-cell supercapacitor.

The impedance of a system can be modeled using equivalent circuit elements, which have impedance characteristics, as shown by Orazem and Tribollet[35]

$$Z_{resistor} = R \qquad (4.13)$$

$$Z_{capacitor} = \frac{1}{j\omega C} \qquad (4.14)$$

and

$$Z_{inductor} = j\omega L \qquad (4.15)$$

where R denotes the resistance, j denotes the imaginary number $\sqrt{-1}$, ω denotes the frequency, C denotes the capacitance, and L denotes the inductance. In supercapacitors, the impedance due to an inductor is often ignored or is considered to contribute only a very small portion of the overall impedance.[34] As can be seen in Equations 4.13 and 4.14, only the impedance of the capacitor has a dependence on frequency (ignoring $Z_{inductor}$, as per the previous sentence). As the frequency decreases, the $Z_{capacitor}$ increases, giving one of the key characteristics of the EIS spectra of supercapacitor systems, as will be described in Section 4.10.2 with Nyquist plots.

The calculation of the total impedance of a system depends on the equivalent circuit used to model the system. The impedance of components in series is given by

$$Z = Z_1 + Z_2 \qquad (4.16)$$

whereas the impedance of components in parallel is calculated using

$$\frac{1}{Z} = \frac{1}{Z_1} + \frac{1}{Z_2} \qquad (4.17)$$

The general forms of these equations are provided here, since the total impedance, and calculation thereof, depend on the equivalent circuit (see Section 4.10.1) proposed for the system, as does the calculation of the capacitance.

Since EIS provides information on the resistive and capacitive components of a system, it is often used to determine the characteristic RC for a supercapacitor system. This is most easily accomplished using a plot of the imaginary part of capacitance (C'') as a function of frequency (described further in Section 4.10.3).

As seen from the above discussion, the physical parameters of a system can be approximated by using circuit elements (resistors and capacitors) and thus, EIS results are often compared with spectra modeled using an equivalent circuit. The fit between the model and the data is used to determine if the circuit adequately describes the physical processes in the system. The fitting of an equivalent circuit model must be used with care and consideration as each element of the equivalent circuit must be chosen to represent a physical process occurring in the real system. An equivalent circuit that fits the data very well but does not explain how each circuit element relates to a process occurring in the system is useless as it does not provide any real knowledge about the system under study. The most common equivalent circuit models for supercapacitors are described in Section 4.10.1.

4.10.1   Equivalent Circuit Modeling

Many different models have been used for supercapacitors to model EIS data, ranging from the simple modified Randles circuit to nested transmission lines, and beyond. The most common equivalent circuits and EIS plot shapes seen in the literature for supercapacitors are described here, but a full discussion of all of the possible equivalent circuits seen in the literature is beyond the scope of this chapter.

One of the most common equivalent circuits for modeling a supercapacitor uses a series connection that includes[29,34] a resistor, which is used to model the equivalent series resistance; and a complex pore impedance (Z_p), which models the frequency dependence of the pore thus giving the 45° phase angle as described by de Levie.[12,13] The Z_p is composed of a number of series-connected (or ladder of) RC circuits, where the resistance in the ladder models the electrolyte resistance in the pore per unit length and the capacitance models the double-layer capacitance at the electrode–electrolyte interface per unit length.[12] Often, this Z_p is modeled as a transmission line, which results in a 45° phase angle.[12] In a real system, there is a variation in the phase angle due to the differences in the pore radius and length, often resulting in a non-45° phase angle.

Some authors also include an inductor in series with the resistor,[29,34] although Buller et al. note that the inductance is not of significant interest in supercapacitors but reduces the error in the resulting spectra.[34] Buller et al. then uses the equivalent circuit developed through comparison with experimental EIS data for a supercapacitor to model the expected output of galvanostatic charge/discharge measurements and calculate EE, with excellent agreement between simulated and measured results, with only a small bias due to self-discharge reactions, which were neglected in the model.[34]

Other equivalent circuits use a constant-phase element (CPE) in place of the capacitor in the modified Randles circuit or other similar circuits. The CPE can be attributed to a number of possibilities, depending on the system under study. CPE effects have been attributed to surface roughness, relating to the work function of different crystal faces, distributed capacitance in the pores, different surface chemistry, differences in pore sizes and diffusion-controlled self-discharge reactions.[29,36–38] Essentially, many of these processes have some rate dependence where either the charging of the double layer or the faradaic reaction of a species cannot keep up with the measurement at high frequencies, but at low rates (low frequencies) there are no such considerations; the CPE influences the Nyquist plot more significantly in the low-frequency region where the vertical line of the capacitor can be seen.[29]

Depending on the system under study, other authors may also include a Warburg element in the circuit to model a diffusion-controlled faradaic reaction;[39] for example, the cation diffusion to a surface in the study by Chun et al.[40] Thus, the wide range of equivalent circuits that are used for fitting the EIS spectra of capacitive materials and system is highlighted. And again, care and consideration must be used when choosing the circuit, as each circuit element must correspond with a physical process occurring on the surface. While increasing the number of circuit elements can result in a better match between experimental and theoretical EIS spectra, a physical process must be assigned to each circuit element in order to gain knowledge about the system.

4.10.2 Nyquist Plots

EIS measurements are plotted in a number of ways. The most common is the Nyquist plot where the imaginary impedance (Z'') is plotted as a function of the real impedance (Z'). With Nyquist plots, the impedance values calculated from high frequencies appear on the left side of the plot, whereas lower frequency data appear on the right. At very high frequencies, the Nyquist plot crosses the x-axis ($Z'' = 0$), and the resistance read from the plot at this point is the minimum resistance (or equivalent series resistance) for the system. At the high frequencies where the ESR is recorded, the capacitance (which is calculated from the Z'') is much lower than the equilibrium capacitance because in porous electrodes only the face of the electrode is electrochemically active.[32] Note that often some negative Z'' results are seen at the highest frequencies, but this is from the induction of the connections and is not used for analysis of supercapacitor systems.

As the frequency drops, the AC signal penetrates deeper into the pores and the capacitance increases, which is responsible for the 45° slope seen at high frequencies.[12,13] This 45° slope is often modeled with the Z_p described above, or a transmission line. The phase angle for a transmission line is 45°;[12] therefore, porous electrodes are expected to have a phase angle of approximately this value. In reality, there is a variation in the phase angle due to differences in the pore radius and length, and often this results in a flattened semicircle with a non-45° phase angle.

The slope of the Nyquist plot then changes at the knee point and the Z'' begins to rapidly increase, giving the (ideally) vertical line indicative of a capacitor. At the lowest frequencies, a capacitor begins to appear as an open-circuit element since there is maximum ion adsorption/full surface charging at all of these low frequencies.[32] Often, the low-frequency line is not truly vertical, and some authors incorporate a CPE to account for this tilt. The physical process modeled by the CPE depends on the system, with explanations including surface roughness, relating to the work function of different crystal faces, distributed capacitance in the pores, different surface chemistry, differences in pores sizes, and diffusion-controlled self-discharge reactions.[29,36–38]

The degree of tilt of the vertical line at low frequencies provides information on the closeness of the system to a true capacitor, where a vertical line indicates that the pores are not influencing the capacitive response of the system (i.e., the system is acting as a pure capacitor).[29] When the tilt is modeled using a CPE, a CPE value close to 1 indicates a good, purely capacitive system.[29]

Often, a semicircle is seen in the Nyquist plot, starting at high frequencies. Typically, the semicircle is depressed (only the top arc of the circle is visible, rather than a true semicircle). Again, the explanation of the physical process, which results in this semicircle, depends on the system under study. Often, the diameter of the semicircle (as measured on the Z' axis) is denoted as R_{ct}; the resistance associated with the charge transfer during a faradaic reaction on the surface.[41–43] For materials with very small (micro or meso) pores, the semicircle may also be ascribed to the adsorption of ions on the surface or diffusion/migration of ions inside pores.[44,45] Depending on the size of the semicircle, the 45° slope described above may or may not be seen.

4.10.3   Other EIS Plots

While Nyquist plots are by far the most common way of presenting EIS data, many researchers also present series capacitance plots, imaginary capacitance plots, complex power plots, series resistance plots and, to a lesser extent, Bode plots. Of these, the two most commonly used plots are those involving series and imaginary capacitance.

In a series capacitance plot, the capacitance is plotted versus frequency. This plot is used particularly often for systems with porous electrodes since the capacitance of these systems is related to the penetration depth into the pores. Typically in supercapacitors, a plateau appears at low frequencies, since at low frequencies the surface is fully charged and there are no ion migration limitations. The value of this plateau is used to provide the maximum capacitance for the electrode/system.[32] At higher frequencies, the capacitance falls as less of the surface/film is charged due to pore effects or ion-migration limitations into the electrode/film. The real part and imaginary part of the capacitance (C' and C'', respectively) are often also plotted as a function of frequency. The plots of C'' are particularly useful as the maximum for this plot occurs at the characteristic frequency (f) related to the time constant ($\tau = 1/2\pi f$), and allows for an easy method to determine this time constant. C'' is calculated using

$$C''(\omega) = \frac{Z'(\omega)}{\omega |Z(\omega)|^2} \tag{4.18}$$

where ω is the frequency.[46]

4.10.4 Capacitance

Depending on the proposed equivalent circuit, the capacitance can be determined from the impedance using the generalized impedance equations in Section 4.10. Capacitance has frequency dependence, particularly in porous electrodes. Since at high frequencies, the capacitance is coming from only the face of the electrode, the value is low. At low frequencies, the whole surface may be charged, resulting in a high capacitance. Using EIS data, it is possible to determine the capacitance at each frequency. Capacitance–frequency profiles are typically plotted with low frequency to the left and high frequency to the right. The capacitance typically climbs as one moves left from high frequency to low, and once the whole surface can be charged by the AC potential (i.e., the penetration depth reaches the bottom of the pore) the capacitance plateaus at its maximum value.

Once the capacitance and resistance have been determined from the EIS model, the energy and power characteristics of the supercapacitor can be determined in the usual way (see Equations 4.4 and 4.8, respectively). Ragone plots can then be constructed for the system.[47]

4.11 Constant Power Discharging

Constant power discharge discharging is one of the less common methods for evaluating a supercapacitor's performance, although constant power discharge can be important practically.[4] With supercapacitors, the potential changes as a function of SOC and since power depends on both potential and current, during constant power discharge experiments both the potential and current change, making the analysis more difficult. Nevertheless, the technique is used, often in order to reduce discharge times (relative to those for galvanostatic discharge).[32] From this technique, energy density, power density, charge, and EE can be calculated.

4.12 Constant Resistance Discharging

Similar to constant power experiments, constant resistance discharging is only rarely used to evaluate supercapacitor performance. In this technique, the supercapacitor is discharged through a load having a constant resistance. Both the voltage and current may be recorded with time.[32] Since the supercapacitor's voltage changes during discharge and the resistance are constant, both the current and power will also change during discharge. Again, this makes the data analysis more complex. Charge may be calculated through the integration of the recorded current over time.

4.13   Potentiostatic Charge/Discharge

In potentiostatic charge/discharge experiments, a potential step is applied to the electrode and the current required to maintain that potential is recorded over time. The technique is conducted in the same way as float current experiments (described in Section 4.15.2) for self-discharge measurements, although the analysis of the resultant current–time profile is slightly different. The charge may be calculated through integration of the current–time profile.[4] Additionally, if the current decays can be fitted with an exponential decay, it suggests that the charging/discharging process is governed by the RC time constant, rather than by ion transport in a pseudocapacitive film.[8] Since, at this point, this technique is not often used to evaluate supercapacitor charging performance, but is more often used to determine float current, the remaining description of this evaluation procedure is described in the float current section (Section 4.15.2).

4.14   Life Cycle Testing

Life cycle testing is often conducted using either cyclic voltammetry or galvanostatic cycling. Alternately, the supercapacitor may be held potentiostatically at the maximum voltage, coupled with periodic cyclic voltammetric or galvanostatic cycling.[26,28] The capacitance is recorded for each charge/discharge cycle and is plotted as a function of the number of cycles. There is some suggestion that the potentiostatic test is the more rigorous testing method since in this test the supercapacitor spends more time at its maximum rated voltage than during the cycling tests.[48] It is suggested that cycling is a gentler life cycle test and produces results that are typically higher than those that would be seen with galvanostatic cycling or in a real supercapacitor system. Degradation increases with increasing temperatures and potentials. Kötz et al. showed an acceleration of the aging process by a factor of 2 for each 10°C increase in temperature or 0.1 V increase in voltage.[47]

4.15   Self-Discharge

Self-discharge is the loss of voltage experienced by a supercapacitor during storage in a charged state. Self-discharge is an important consideration for supercapacitors because often supercapacitors experience higher rates of self-discharge than that of batteries. Self-discharge causes a drop in the energy and power of a supercapacitor since energy and power both are related to voltage (Equations 4.4 and 4.8).

Self-discharge experiments are typically conducted in two ways: open-circuit potential decay measurements and float (or leakage) current measurements. As described in Section 4.16.1, the potential decay measurements provide a method to elucidate the mechanism or rate-determining step for the faradaic reaction causing self-discharge. Float current measurements can provide information on the charge redistribution in a porous electrode.

4.16 Potential Decay

Potential decay measurements are conducted by first charging the electrode or supercapacitor to some desired voltage. This desired voltage may or may not be held for some time; a holding step of this potential is used ensure the whole surface is at the same potential, and thereby reducing/eliminating charge redistribution. Finally, the system is set on open circuit and the potential is recorded as a function of time, typically for tens of hours. Potential decay measurements may be conducted in either a two-electrode or three-electrode setup. A three-electrode configuration results in more useful information because it allows the researcher to begin to identify, using the models described in the next section, the rate-determining step of the self-discharge reaction occurring on the working electrode. Each supercapacitor electrode can be examined independently. In a two-electrode configuration, the overall rate of self-discharge is observed, giving a realistic measure of the self-discharge behavior of the system; but it is not known which electrode is experiencing the greatest degree of self-discharge, or what the reaction for that electrode may be.

4.16.1 Conway Models and the Rate-Determining Step of Self-Discharge Reactions

Self-discharge profiles may be used to elucidate the rate-determining step of the mechanism of the faradaic self-discharge reaction. Conway et al. proposed three mathematical models that determine whether the self-discharge is under activation control, diffusion control, or caused by ohmic leakage.[16] In this way, one may focus on the search for the identity of the self-discharge reaction and discard other possible mechanisms. For instance, if the self-discharge reaction is determined to be diffusion controlled, then it is unlikely that this reaction is electrolyte decomposition (although decomposition of electrolyte impurities is still possible) or the reaction of a surface species. However, if the self-discharge is activation controlled, then one may discard the possibility of self-discharge due to low-concentration impurities.

4.16.1.1   Activation-Controlled Mechanism

In activation-controlled reactions, the reacting species must be at high concentrations or attached to the electrode (i.e., the reaction does not depend on diffusion). Examples of this type of self-discharge are decomposition of the electrolyte solvent or the reaction of a surface functionality on a carbon electrode. Conway showed that an activation-controlled self-discharge reaction should follow the equation:[16]

$$V_t = -\frac{RT}{aF} \ln \frac{aFi_o}{RTC} - \frac{RT}{aF} \ln \left[t + \frac{C\tau}{i_o} \right] \tag{4.19}$$

where V_t is the voltage at any time, t, during the self-discharge, V_i is the initial voltage to which the system or electrode was charged, R is the gas constant, T is temperature, a is the transfer coefficient, F is faraday's constant, i_o is the exchange current density, C is the interfacial capacitance, and τ is the integration constant. An activation-controlled self-discharge would then result in a linear change in potential as a function of the natural logarithm of time, after some plateau (τ) given by[16]

$$\tau = \left[\frac{RT}{aF} \right] \frac{i_o}{i_i} \tag{4.20}$$

where i_i is the initial current upon polarization. The slope of the plot is the negative of the Tafel slope for the self-discharge reaction.

In the situation where the activation-controlled process includes an adsorbed species on the surface (such as the case of underpotential deposition of hydrogen during H_2 evolution), there will be two slopes, corresponding to the reaction at low and high surface coverage (θ) of the adsorbed species.[16]

As can be seen from Equation 4.19, the slope should be independent of the initial charging potential, and if a comparison is made of the system when charged to different potentials, all slopes should be the same at any particular potential. If this condition is satisfied, then the slope of the self-discharge profile can be used to determine the Tafel slope for the self-discharge reaction. However, if there is charge redistribution in the system due to incomplete charging of a pseudocapacitive film or porous electrode (described in more detail in Section 4.16.1.4), then the slope will depend on the initial potential. Thus, a dependence of slope on initial potential indicates the possibility of the presence of charge redistribution in the system, and therefore, the slope of the self-discharge plot cannot be used for Tafel slope calculations.

4.16.1.2   Diffusion-Controlled Mechanism

Diffusion was also modeled as a rate-determining step for self-discharge reactions,[16] as would be the case when self-discharge is caused by the faradaic reaction of a low-concentration impurity in the supercapacitor (such as metal contaminants in aqueous supercapacitors). As discussed later in this section, this model also applies where there is a diffusion of oxidation states through an incompletely charged or discharged metal oxide film.

Diffusion-controlled self-discharge is described by:[16]

$$C(V_i - V_t) = 2zFAD^{1/2}\pi^{-1/2}c_0t^{1/2} \qquad (4.21)$$

where z is the charge on the diffusion species, A is the electrode area, D is the diffusion constant of the reacting species, and c_0 is the initial concentration. A plot of V_t-V_i as a function of $t^{1/2}$ results in a linear slope if the self-discharge reaction is diffusion controlled.

Of particular interest for metal oxide pseudocapacitive materials is a diffusion-controlled self-discharge or potential recovery exhibited such as that seen in ruthenium oxide films upon stopping the charging or discharging process.[16] During charging or discharging, the species at the film surface changes the oxidation state first, while the species deeper in the film lags behind. If the film is not fully charged or discharged, then, when the charge or discharge is stopped, the oxidation states of the species in the film come to equilibrium and this process is controlled by the diffusion of the oxidation states in the film.

4.16.1.3 Ohmic Leakage between Two Plates

Conway also modeled self-discharge due to an ohmic leakage between the two supercapacitor electrodes, (i.e., an electrical connection between the plates)[16]. The self-discharge is then governed by:

$$\frac{V}{V_i} = \exp\frac{-t}{RC} \qquad (4.22)$$

and a plot of the log potential as a function of time results in a linearly decreasing profile.

4.16.1.4 Charge Redistribution

During charging or discharging of highly porous electrodes, the potential at the tip of the pores changes faster than the potential of the surface deeper in the pores. This is due to the distributed resistance down the pore, as discussed in detail by de Levie.[12,13] Thus, when a highly porous electrode is charged, there is a distribution of potential down the pores of the electrode. When charging or discharging is stopped, the charge moves through the pore in order to equalize this potential, called charge redistribution. Since the electrode potential is measured at the tip of the pore, the movement of charge deeper into the pore after charging appears as a drop in voltage (self-discharge).

Self-discharge profiles based on pure charge distribution have been modeled using a transmission line hardware circuit.[49] It was shown that the apparent self-discharge that resulted when a system undergoes charge redistribution is similar to that seen with a faradaic self-discharge reaction; that is, the plot of self-discharge voltage as a function of log time exhibits an initial plateau in potential followed by a linear decay, in both situations. Therefore, for porous electrodes, the existence of this shape in the self-discharge plot is not characteristic of an activation-controlled faradaic reaction, and caution should be used when using this shape to state that an activation-controlled faradaic reaction is responsible for the self-discharge. For porous electrodes, this shape may indicate that the self-discharge seen in the system is due to charge redistribution, rather than a faradaic reaction.

One characteristic that can differentiate the two possible mechanisms is the dependence of the slope on the initial charging potential. Self-discharge caused purely by an activation-controlled faradaic reaction should have a slope at each potential which is independent of the initial charging voltage of the system, as described earlier. Self-discharge which is, at least partially, caused by charge redistribution will have a slope dependence on initial potential. Therefore, comparing the self-discharge profiles (plotted as a function of log time) recorded for the system charged to different initial potentials can give a method to determine if charge redistribution is influencing the self-discharge in the supercapacitor.

Since charge redistribution is a result of a distributed potential down the pore caused by the pore solution resistance, the diameter, length, and shape of the pore all influence the solution resistance and therefore the charge distribution.[14,49] As expected, longer or narrower pores set up a large potential distribution down their length, resulting in more charge redistribution. Pore shape results in much more complex charge redistribution (and self-discharge) profiles. The existence of bottlenecks in the pores can drastically impact the charge redistribution and self-discharge profile shape, and therefore should be avoided when possible.

4.17   Float Current Measurements

The self-discharge current is the current that flows through the faradaic reaction causing the self-discharge. In a float current (also called leakage current and leak current) measurement, the voltage of the electrode or supercapacitor is stepped to the desired value and then the current (I_f) that is recorded, which is required to maintain the system at this voltage is recorded. The resulting current is monitored, ideally, for several hours. Float current measurements can be conducted in two-electrode or three-electrode configurations, although typically they are two-electrode (full cell) measurements as this gives a more realistic measure of the float current of a real supercapacitor. Nevertheless, three-electrode configurations can be useful for identifying the specific self-discharge reaction occurring on each individual supercapacitor electrode. The term "leakage current" is also used to mean current that leaks between electrodes through a short circuit in the cell.[50] Thus, for clarity, the term float current is used in this chapter.

The float current is plotted as a function of time. In planar double-layer electrodes, the current follows the typical potentiostatic current decay in microseconds, followed by a steady-state current. More typically with real supercapacitor electrodes, the current decay may require tens to thousands of seconds[1,8], and is followed by a steady-state current (Figure 4.7). The float current decay can be associated with the amount of charge redistribution in the film or porous electrode and the steady-state current is equal to the self-discharge current.[51]

The decay in float current early in the measurement is associated with the current required to fully charge the surface to the desired potential and the shape of the decay can provide qualitative information about the

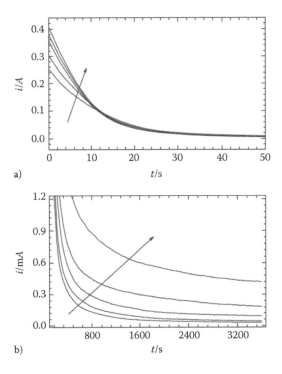

FIGURE 4.7
Float current measurements for Spectracarb 2225 in 5 M H₂SO₄ following polarization to various initial potentials of 0.8 V, 0.9 V, 1.0 V, and 1.05 V, where the arrow shows an increasing polarization potential. (From Niu, J. et al., *J. Power Sources*, 135, 332–343, 2004. With permission.)

ESR of the system (particularly the solution resistance in the pores or the resistivity of a pseudocapacitive film).[51] When the system resistance is low (materials with wide pores or no film migration effects), then more of the surface is electrochemically active immediately, relative to high-resistance situations, leading to more initial charging of the surface and resulting in a high initial float current.[1] Additionally, low resistivity means that ion migration into the pores or through the film is very fast, and the float current decay is relatively rapid. Thus, for a low-resistance system, the float current profile is tall with a rapid decay. When the resistance in the system is high, less of the surface/film is active initially, resulting in a low initial float current.[1] And, since ion migration is slow in these systems, the float current decays slowly as it takes longer to charge the whole surface/film. Thus, with high-resistance systems, the float current is low and broad (requiring longer times). Long times of current decay (and therefore higher float charges as calculated by integration under this decay) indicate that a large portion of the surface is not fully charged and suggests that this electrode will undergo charge redistribution after charging in the typical way.

Since the steady-state current is equivalent to the current flowing through the faradaic self-discharge reaction, it is theoretically possible to directly compare the current density between systems to determine which has the greatest degree of self-discharge. Caution must be taken to ensure that the systems are normalized to an appropriate measure (i.e., electroactive area when possible) when doing these direct comparisons since the self-discharge reaction, and therefore float current, will be related to the electrode surface area, distributed potential profile in the pores, issues with assembly and packaging with full supercapacitors, etc. Therefore, self-discharge comparisons between systems must be done with caution and in replicates. It is easier to directly compare float current measurements when the same electrode is examined under different conditions (e.g., different electrolyte concentrations or different polarization potentials).

Different initial polarization potentials are of particular interest since these float current measurements can indicate when the self-discharge reactions begin or when there is a change in self-discharge mechanism.[1] To identify the onset of self-discharge reactions or a change in mechanism, the float charge (Q_f, calculated through the integration of the current vs. time curve) is plotted as a function of polarization potential. A change in slope in this profile indicates that there is a change is self-discharge mechanism (Figure 4.8).[1] The float charge increases with potential because (1) charging the double layer to a higher potential requires more charge to be placed on the surface, since the potential is directly related to the charge separation across the electrode–electrolyte interface; and (2) at higher potentials, in the presence of a faradaic reaction, more current will flow for that reaction, and therefore more charge will pass into that reaction.

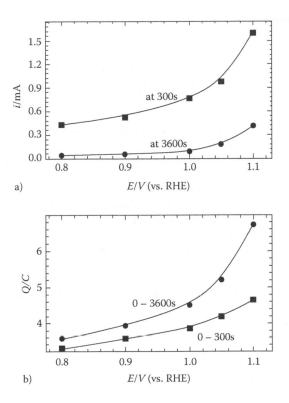

FIGURE 4.8
Float current a) and float charge b) as a function of polarizing potential for Spectracarb 2225 in 5 M H$_2$SO$_4$. (From Niu, J. et al., *J. Power Sources*, 135, 332–343, 2004. With permission.)

4.18 Conclusion

As described in this chapter, there are many techniques used to characterize supercapacitor materials and system, but the most common are cyclic voltammetry, galvanostatic charge/discharge, and EIS. Cyclic voltammetry, and galvanostatic charge/discharge experiments can provide key supercapacitor characteristics such as stable voltage window, capacitance, capacity, energy, power, and coulombic and energy efficiencies. EIS is used to quantify the resistive and capacitive components of the system as well as provide a means to model the physical processes at the surface. The electrolyte resistance in pores and the resistance to ion migration in pseudocapacitive films can significantly impact a supercapacitor's performance, and their impact on supercapacitor characteristics was outlined here.

References

1. Niu J, Pell WG, Conway BE. Requirements for performance characterization of C double-layer supercapacitors: Applications to a high specific-area C-cloth material. *J Power Sources* 2006; 156: 725–740.
2. Pell WG, Conway BE, Adams WA, et al. Electrochemical efficiency in multiple discharge/recharge cycling of supercapacitors in hybrid EV applications. *J Power Sources* 1999; 80: 134–141.
3. Winter M, Brodd RJ. What are batteries, fuel cells, and supercapacitors?. *Chem Rev* 2004; 104: 4245–4269.
4. Conway BE. *Electrochemical Supercapacitors: Scientific Fundamentals and Technological Applications.* New York, Kluwer-Plenum Publ. Co, 1999.
5. Frackowiak E. Carbon materials for supercapacitor application. *Phys Chem Chem Phys* 2007; 9: 1774–1785.
6. McKeown DA, Hagans PL, Carette LPL, et al. Structure of hydrous ruthenium oxides: Implications for charge storage. *J Phys Chem B* 1999; 103: 4825–4832.
7. Sivakkumar SR, Ko JM, Kim DY, et al. Performance evaluation of CNT/polypyrrole/MnO2 composite electrodes for electrochemical capacitors. *Electrochim Acta* 2007; 52: 7377–7385.
8. Nyström G, Strømme M, Sjödin M, et al. Rapid potential step charging of paper-based polypyrrole energy storage devices. *Electrochim Acta* 2012; 70: 91–97.
9. Grahame DC. The electrical double layer and the theory of electrocapillarity. *Chem. Rev.* 1947; 41, 441–501.
10. Fedorov MV, Kornyshev AA. Towards understanding the structure and capacitance of electrical double layer in ionic liquids. *Electrochim Acta* 2008; 53: 6835–6840.
11. Feng G, Huang J, Sumpter BG, et al. A "counter-charge layer in generalized solvents" framework for electrical double layers in neat and hybrid ionic liquid electrolytes. *Phys Chem Chem Phys* 2011; 13: 14723–14734.
12. de Levie R. Porous electrodes in electrolyte solutions. *Electrochim Acta* 1963; 8: 751–780.
13. de Levie R. Porous electrodes in electrolyte solutions IV. *Electrochim Acta* 1964; 9: 1231–1245.
14. Black JM, Andreas HA. Pore shape affects spontaneous charge redistribution in small pores. *J Phys Chem C* 2010; 114: 12030–12038.
15. Itagaki M, Hatada Y, Shitanda I, et al. Complex impedance spectra of porous electrode with fractal structure. *Electrochim Acta* 2010; 55: 6255–6262.
16. Conway BE, Pell WG, Liu T. Diagnostic analyses for mechanisms of self-discharge of electrochemical capacitors and batteries. *J Power Sources* 1997; 65: 53–59.
17. Toupin M, Brousse T, Bélanger D. Charge storage mechanism of MnO2 electrode used in aqueous electrochemical capacitor. *Chem Mater* 2004; 16: 3184–3190.
18. Long JW, Swider KE, Merzbacher CI, et al. Voltammetric characterization of ruthenium oxide-based aerogels and other RuO2 solids: The nature of capacitance in nanostructured materials. *Langmuir* 1999; 15: 780–785.
19. Frackowiak E, Béguin F. Carbon materials for the electrochemical storage of energy in capacitors. *Carbon* 2001; 39: 937–950.

20. Ryoo R, Joo SH, Kruk M, et al. Ordered mesoporous carbons. *Adv Mater* 2001; 13: 677–681.
21. Kyotani T. Control of pore structure in carbon. *Carbon* 2000; 38: 269–286.
22. Kyotani T, Ma Z, Tomita A. Template synthesis of novel porous carbons using various types of zeolites. *Carbon* 2003; 41: 1451–1459.
23. Andreas HA, Conway BE. Examination of the double-layer capacitance of an high specific-area C-cloth electrode as titrated from acidic to alkaline pHs. *Electrochim Acta* 2006; 51: 6510–6520.
24. Algharaibeh Z, Liu X, Pickup PG. An asymmetric anthraquinone-modified carbon/ruthenium oxide supercapacitor. *J Power Sources* 2009; 187: 640–643.
25. Bélanger D, Ren X, Davey J, et al. Characterization and long-term performance of polyaniline-based electrochemical capacitors. *J Electrochem Soc* 2000; 147: 2923–2929.
26. Cericola D, Ruch PW, Foelske-Schmitz A, et al. Effect of water on the aging of activated carbon based electrochemical double layer capacitors during constant voltage load tests. *Int J Electrochem Sci* 2011; 6: 988–996.
27. Ishimoto S, Asakawa Y, Shinya M, et al. Degradation responses of activated-carbon-based EDLCs for higher voltage operation and their factors. *J Electrochem Soc* 2009; 156: A563–A571.
28. Bittner AM, Zhu M, Yang Y, et al. Ageing of electrochemical double layer capacitors. *J Power Sources* 2012; 203: 262–273.
29. Kim S, Choi W. Selection criteria for supercapacitors based on performance evaluations. *J Power Electron* 2012; 12: 223–231.
30. Smith AJ, Burns JC, Trussler S, et al. Precision measurements of the coulombic efficiency of lithium-ion batteries and of electrode materials for lithium-ion batteries. *J Electrochem Soc* 2010; 157: A196–A202.
31. Izadi-Najafabadi A, Tan DTH, Madden JD. Towards high power polypyrrole/carbon capacitors. *Synth Met* 2005; 152: 129–132.
32. Arulepp M, Leis J, Latt M, et al. The advanced carbide-derived carbon based supercapacitor. *J Power Sources* 2006; 162: 1460–1466.
33. Conway BE, Pell WG. Power limitations of supercapacitor operation associated with resistance and capacitance distribution in porous electrode devices. *J Power Sources* 2002; 105: 169–181.
34. Buller S, Karden E, Kok D, et al. Modeling the dynamic behavior of super-capacitors using impedance spectroscopy. *IEEE Trans Ind Appl* 2002; 38: 1622–1626.
35. Orazem ME, Tribollet B. *Electrochemical Impedance Spectroscopy*. New Jersey: John Wiley and Sons, Inc., 2008.
36. Song H, Jung Y, Lee K, et al. Electrochemical impedance spectroscopy of porous electrodes: the effect of pore size distribution. *Electrochim Acta* 1999; 44: 3513–3519.
37. Kerner Z, Pajkossy T. Impedance of rough capacitive electrodes: The role of surface disorder. *J Electroanal Chem* 1998; 448: 139–142.
38. Pajkossy T. Capacitance dispersion on solid electrodes: Anion adsorption studies on gold single crystal electrodes. *Solid State Ionics* 1997; 94: 123–129.
39. Gassa LM, Mishima HT, deMishima BAL, et al. An electrochemical impedance spectroscopy study of electrodeposited manganese oxide films in borate buffers. *Electrochim Acta* 1997; 42: 1717–1723.

40. Chun S, Pyun S, Lee G. A study on mechanism of charging/discharging at amorphous manganese oxide electrode in 0.1 M Na2SO4 solution. *Electrochim Acta* 2006; 51: 6479–6486.
41. Laheäär A, Jänes A, Lust E. Lithium bis(oxalato)borate as an electrolyte for micromesoporous carbide-derived carbon based supercapacitors. *J Electroanal Chem* 2012; 669: 67–72.
42. Laheäär A, Peikolainen A, Koel M, et al. Comparison of carbon aerogel and carbide-derived carbon as electrode materials for non-aqueous supercapacitors with high performance. *J Solid State Electrochem* 2012; 16: 2717–2722.
43. Li H, Wang J, Chu Q, et al. Theoretical and experimental specific capacitance of polyaniline in sulfuric acid. *J Power Sources* 2009; 190: 578–586.
44. Eskusson J, Jänes A, Kikas A, et al. Physical and electrochemical characteristics of supercapacitors based on carbide derived carbon electrodes in aqueous electrolytes. *J Power Sources* 2011; 196: 4109–4116.
45. Thomberg T, Jänes A, Lust E. Energy and power performance of electrochemical double-layer capacitors based on molybdenum carbide derived carbon. *Electrochim Acta* 2010; 55: 3138–3143.
46. Taberna PL, Simon P, Fauvarque JF. Electrochemical characteristics and impedance spectroscopy studies of carbon-carbon supercapacitors. *J Electrochem Soc* 2003; 150: A292–A300.
47. Kötz R, Hahn M, Gallay R. Temperature behavior and impedance fundamentals of supercapacitors. *J Power Sources* 2006; 154: 550–555.
48. Weingarth D, Foelske-Schmitz A, Kötz R. Cycle versus voltage hold - Which is the better stability test for electrochemical double layer capacitors?. *J Power Sources* 2013; 225: 84–88.
49. Black J, Andreas HA. Effects of charge redistribution on self-discharge of electrochemical capacitors. *Electrochim Acta* 2009; 54: 3568–3574.
50. Diab Y, Venet P, Gualous H, et al. Self-discharge characterization and modeling of electrochemical capacitor used for power electronics applications. *IEEE Trans Power Electron* 2009; 2: 510–517.
51. Niu J, Conway BE, Pell WG. Comparative studies of self-discharge by potential decay and float-current measurements at C double-layer capacitor and battery electrodes. *J Power Sources* 2004; 135: 332–343.

5

Nanostructured Metal Oxides for Supercapacitor Applications

P. Ragupathy, R. Ranjusha, and Roshny Jagan

CONTENTS

5.1   Introduction

Supercapacitors can be classified into two categories, namely, electrical double-layer capacitors (EDLCs) with carbon materials as electrodes and pseudocapacitors with transition metal oxides or conducting polymers as electrodes.[1-3] However, one major limitation of carbon-based EDLCs is the lower specific energy density. Most of the commercial products available in the market have a specific energy density lower than 10 Whkg^{-1}. This specific energy density value is still relatively lower than the lowest figures for batteries, that is, 35–40 Whkg^{-1} (lead acid battery). Metal oxides provide an alternative solution as an electrode material because of their high specific capacitance (SC; farad per gram) at low resistance, thereby making it easier to construct high energy and high power density supercapacitors. Several transition metal oxides with various oxidation states represent attractive materials for supercapacitor electrodes owing to their excellent structural stability and high SC. For the transition metal oxides, the reversible redox reactions are primarily responsible for the energy storage in addition to the electric double-layer storage. This makes metal oxides with pseudoca-pacitance a predominant part in charge storage process in supercapacitors. There have been plenty of researches on utilization of these metal oxides in different nanostructure forms to enhance the overall performance of the supercapacitors. The metal oxides that have been explored as supercapacitor electrodes, including RuO_2, NiO, MnO_2, Co_2O_3, IrO_2, FeO, TiO_2, SnO_2, and V_2O_5, that exhibit high SC values.[4,5]

As we know, nanotechnology can manipulate particles on both atomic and molecular scales, the nanometer products such as metal oxides used in supercapacitors can be processed by two techniques: Top-down or Bottom-up approach (Figure 5.1). The top-down approach starts with a bulk material, which breaks down into smaller fragments. So this approach makes use of larger (initial) structures, which is controlled externally in the process-ing of nanostructures, for example, ball milling or attrition and lithography (etching through a mask).

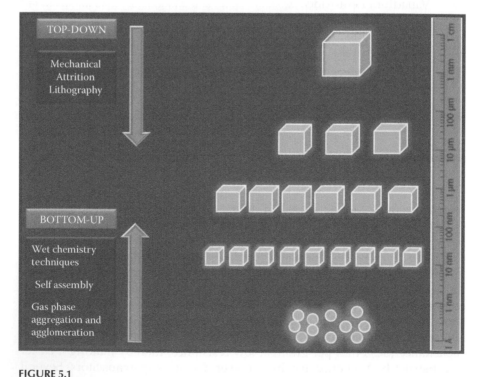

FIGURE 5.1
Schematic representation of the "Top-down" and "Bottom-up" approaches of nanomaterials with the different techniques that are used.

The bottom-up approach is a buildup of material from atomic scale to nanoregime (nanoscale): which includes processes like (1) self-assembly and (2) liquid–solid transformation. The liquid–solid transformation approach includes (1) coprecipitation methods, (2) sol-gel processing, (3) microemulsion technique, and (4) solvothermal methods. The coprecipitation method involves dissolving a salt precursor (chloride, nitrate, etc.) in water (or other solvent) to precipitate the oxohydroxide form using a base. Sol-gel processing method prepares metal oxides via hydrolysis of precursors. Microemulsion technique is based on the formation of micelles (nanoreaction vessels) in a mixture containing water, surfactant, and oil. In solvothermal methods, the metal complexes are decomposed thermally in an inert atmosphere under pressure. Remarkable achievements are made in the field of supercapacitor performance by moving from bulk to nanostructured materials.[6–8]

Hybrid energy storage devices have also received enormous interest as they possess very high energy and power densities. They combine a lithium ion (Li-ion) battery with a supercapacitor to achieve maximum storage capacity. In general, hybrid Li-ion supercapacitors are similar to conventional Li-ion battery except the charge storage is at the surface of the electrodes instead of within the electrodes. Such type of hybrid Li-ion supercapacitors are also been discussed in this chapter.

5.2 Ruthenium Oxide

In 1971, the pseudocapacitive behavior of ruthenium oxide exhibiting rectangular current–voltage characteristic for an ideal capacitor was recognized. The crystal structure of ruthenium oxide (RuO_2) is shown in Figure 5.2. Each Ru atom is octahedrally surrounded by six oxygen atoms through edge-sharing RuO_6 octahedrons running along the four fold [001] axis, and neighboring octahedrons in the tetragonal direction cause the length of the shared O-O edges (along [110]) to be 2.47 Å and two edges along [001] are elongated to 3.11 Å, while the distance of the remaining eight O-O edges is 2.78 Å.[9]

RuO_2 in both crystalline and amorphous forms has become appealing as electrode material for supercapacitors due to its unique characteristics such as metallic conductivity,[10] high chemical and thermal stability,[11] and electrochemical redox properties.[12,13] Particularly, hydrated RuO_2 has been extensively studied for electrochemical capacitors due to solid-state pseudo-faradaic reaction exhibiting high theoretical SC of 1358 F g^{-1}. A significant breakthrough was made by Zheng et al. who reported that amorphous RuO_2 obtained by sol-gel process exhibits a higher SC of 720 F g^{-1} than crystalline RuO_2.[13] The superior performance was attributed to hydrous surface layers,

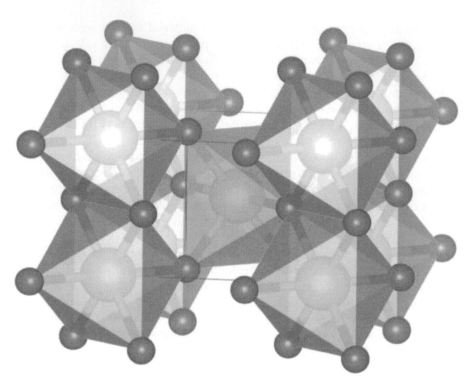

FIGURE 5.2
Bulk crystal structure of RuO_2. Tetragonal z-axis is oriented along the vertical direction. (From Ozoliņš, V. et al., *Accounts of Chemical Research*, 46, 1084–1093, 2013. With permission.)

which facilitate mixed protonic–electronic conductivity. However, the charge storage capability of amorphous RuO_2 decreases at a high scan rate due to proton depletion during charge–discharge cycling.[14]

Much attention has been paid to improve the rate performance by preparing small-sized particles with conductive carbons and porous structures. For example, the ordered, 3D porous architecture is always beneficial due to easy penetration of the electrolyte into whole oxide matrix. Moreover, these 3D mesoporous architectures will minimize the equivalent series resistance by complete utilization of oxide through facile electrolyte permeation, fast proton exchange, and the metallic conductivity. Recently, Hu et al. have made remarkable enhancement in terms of capacitance, energy, and power density by designing 3D mesoporous architecture of $RuO_2.xH_2O$ nanotubular arrayed electrodes.[15] The surface of the opening was rough due to adherent attack of oxide particulates from the bottom of pores, resulting in the formation of these nanotubes with the outer diameter of 200 ± 20 nm as shown in Figure 5.3. The SC of nanotubular arrayed electrodes was increased from 740 to 1300 F g^{-1} by annealing the nanotubes at 200°C for 2 hours. This ultrahigh SC value was mainly due to full utilization of oxides through thin (40 nm) and uniform walls of $RuO_2.xH_2O$ nanoarrays.

FIGURE 5.3
SEM surface images of $RuO_2 \cdot xH_2O$ NTs arrayed electrode. (From Hu, C.-C. et al., *Nano Letters*, 6, 2690–2695, 2006. With permission.)

Brumbach et al. showed the nanostructure RuO_2 obtained through a high-temperature molecular templating route for supercapacitor applications.[16] Figure 5.4 shows the cyclic voltammograms of untemplated RuO_2 and templated RuO_2, revealing that the templated RuO_2 yielded four times higher current than the untemplated RuO_2. The enhanced current from nanostructured RuO_2 was attributed to high utilization and easy permeation of the electrolyte into RuO_2.

Zhang et al.[17] have described the nanotubular ruthenium oxides obtained using magnetite as a morphology sacrificial template. The advantage of this method is that manganese oxides dissolve away during the formation of tubular RuO_2. Electrochemical capacitor study of nanotubes RuO_2 was carried out on two different electrolytes such as H_2SO_4 and Na_2SO_4. The SC values were found to be 861 F g^{-1} and 313 F g^{-1} for H_2SO_4 and Na_2SO_4, respectively, from the charge–discharge profiles (Figure 5.5). The duration of charging and discharging process was almost equal for each electrode, implying the high columbic efficiency of charge–discharge cycling. The variation in the capacitance values was mainly due to sufficient proton transport pathways available in H_2SO_4.

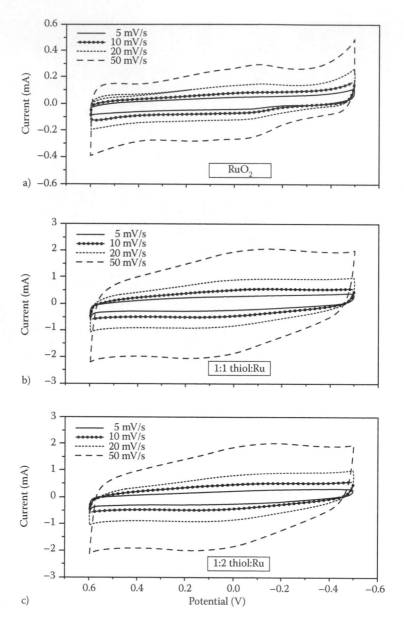

FIGURE 5.4

Cyclic voltammetry of ruthenium oxide thin films on titanium are shown for a) thermally prepared RuO_2 and RuO_2 formed with b) 1:1 and c) 1:2 molar additions of thiol relative to Ru in the precursor solution. The electrolyte solution was 1 M H_2SO_4 and potentials are referenced to $Hg/HgSO_4$. Three voltammograms are shown for each scan rate. Scan rates were 5 (straight-line), 10 (dotted line with strikeout), 20 (small dash lines), and 50 (large dash lines) mV/s. (From Brumbach, M.T. et al., *ACS Applied Materials & Interfaces*, 2, 778–787, 2010. With permission.)

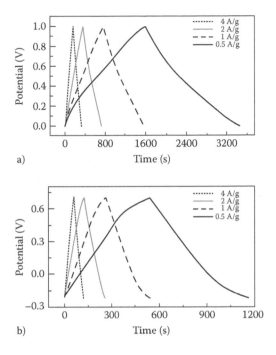

a)

b)

FIGURE 5.5

Charge–discharge curves of nanotubular RuOx·nH$_2$O electrode in H$_2$SO$_4$ a) and in Na$_2$SO$_4$ b) at different current densities. (From Zhang, J. et al., *Journal of Physical Chemistry C*, 114, 13608–13613, 2010. With permission.)

5.3 Manganese Oxide

Despite the high capacitance and long cycle life of ruthenium-based electrodes for supercapacitors, ruthenium is still hampered from its commercialization due to its cost and toxic nature. Hence, alternative metal oxides, such as MnO$_2$, NiO, and Co$_3$O$_4$, are being studied. Among the various metal oxides investigated so far, MnO$_2$ has been appealing as an active material for electrochemical supercapacitors due to its low cost, availability in abundance, and environmental-friendly nature. In 1999, Lee and Goodenough published a pioneering work on the supercapacitor of MnO$_2$ in aqueous electrolyte.[18] Faradaic reactions occurring on the surface of MnO$_2$ and in the bulk MnO$_2$ are the two main mechanisms involved in the charge storage process. The surface faradaic processes are mainly due to surface absorption/desorption of electrolyte cations.

$$\left(\text{MnO}_2\right)_{\text{surface}} + \text{C}^+ + \text{e}^- \leftrightarrow \left(\text{MnOOC}\right)_{\text{surface}} \tag{5.1}$$

where $C^+ = H^+$, Li^+, Na^+, and K^+, while the bulk faradaic process involves the intercalation/deintercalation of cations from the electrolyte.

$$MnO_2 + C^+ + e^- \leftrightarrow MnOOC \tag{5.2}$$

Generally, hydrated MnO_2 exhibits capacitance of 100–200 F g^{-1}, which is much lower than that of RuO_2. MnO_2 exists in different crystallographic structures, namely α, β, γ, δ, and λ. Each MnO_2 crystal structure consists of basic MnO_6 octahedron units, which are linked in different ways to produce different crystallographic forms.[19] The 1D, 2D, and 3D tunnel structures are built by different ways of sharing the vertices and edges of MnO_6 octahedron units.[20] The different crystallographic forms are delineated by the size of the tunnel formed with the number of octahedron subunits ($n \times m$). The structures are schematically depicted in Figure 5.6, and the size and type of tunnels are presented in Table 5.1.

α-MnO_2 (Figure 5.6-α) is composed of double chains of edge-sharing MnO_6 octahedrons, which are linked at corners to form 1D (2×2) and (1×1)

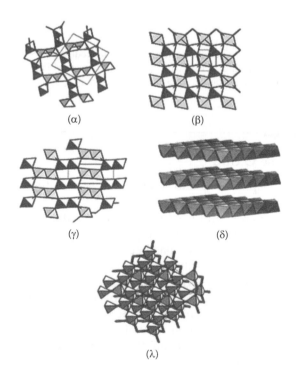

(α)   (β)

(γ)   (δ)

(λ)

FIGURE 5.6
Crystal structures of α-, β-, γ-, δ-, and λ-MnO_2. (From Devaraj, S. and N. Munichandraiah, *Journal of the Electrochemical Society*, 154, A80–A88, 2007. With permission.)

TABLE 5.1

Different Crystallographic Forms of MnO_2

Crystallographic Form	Tunnel	Size/Å
α	(1×1), (2×2)	1.89, 4.6
β	(1×1)	1.89
γ	(1×1), (1×2)	1.89, 2.3
δ	Interlayer distance	7.0

tunnels that extend in a direction parallel to the *c*-axis of the tetragonal unit cell. A small amount of cations such as Li^+, Na^+, K^+, NH_4^+, Ba^{2+}, or H_3O^+ is required to stabilize the (2×2) tunnels in the formation of α-MnO_2.[21]

β-MnO_2 (Figure 5.6-β) is composed of single strands of edge-sharing MnO_6 octahedrons to form a 1D (1×1) tunnel. β-MnO_2 cannot accommodate cations due to the narrow (1×1) tunnel size (~1.89 Å). γ-MnO_2 (Figure 5.6-γ) has a random intergrowth of ramsdellite (1×2) and pyrolusite (1×1) domains.[22] δ-MnO_2 (Figure 5.6-δ) has a 2D layered structure with an interlayer separation of ~7 Å.[23] It has a substantial amount of water and stabilizing cations such as Na+ or K+ between the layers. λ-MnO_2 (Figure 5.6-λ) has a 3D spinel structure.[24] Since manganese-based metal oxides have low conductivity, a significant amount of conductive materials such as carbon is very essential to achieve optimum performance of metal oxides.[25] With an aim to understand the effect of carbon on supercapacitors' performance, cyclic voltammograms and charge–discharge studies were performed at different quantities of carbon in 0.1 M Na_2SO_4 at 20 mV/s, and the data are shown in Figure 5.7. It was observed that as the carbon content increased from 5 to 20 wt%, voltammograms changed from oval to rectangular capacitive shape. However, the shape of cyclic voltammograms remained the same even after 40 wt%.

Thus, 20 wt% of carbon is ideal to create an electric path between the oxide particles. Variation in SC with carbon content is shown in Figure 5.7h. There is a steady increase in SC with an increase in the carbon upto 20 wt%, and thereafter, SC remains nearly constant. It has been widely accepted that the hydrated MnO_2 powder electrodes exhibit ideal capacitive behavior in aqueous electrolytes. Hence, amorphous hydrated MnO_2 is prepared by various methods using inorganic and organic reducing agents. Typically, $KMnO_4$ is being reduced by $MnSO_4$, potassium borohydride, sodium hypophosphite, sodium dithionite, and hydrochloric acid.[26–28]

In addition to these aqueous-based methods, amorphous MnO_2 is also obtained by reducing $KMnO_4$ with surfactant and H_2O/CCl_4 and ferrocene/chloroform solution.[29,30]

The as-prepared hydrated MnO_2, generally, is nanocrystalline or amorphous in nature. The amorphous characteristic of MnO_2 was maintained up to 300°C. However, as the temperature increased to 400°C and above, the hydrated MnO_2 transformed into highly crystalline Mn_2O_3. For instance, Jeong and Manthiram studied the electrochemical redox

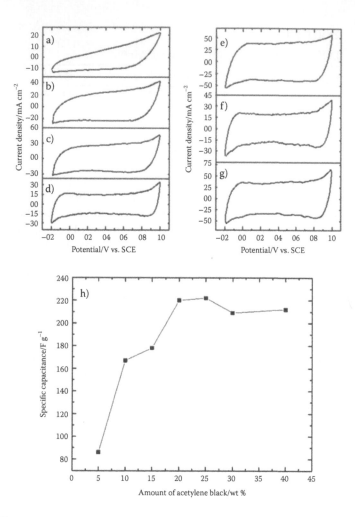

FIGURE 5.7
Cyclic voltammograms of the as-prepared MnO_2 recorded in 0.1 M Na_2SO_4 at 20 mV s^{-1} with carbon content 5 a), 10 b), 15 c), 20 d), 25 e), 30 f), and 40 wt% g) and dependence of SC on the amount of acetylene black in the electrode material h).

capacitance of nanocrystalline MnO_2 prepared by the reduction of aqueous $KMnO_4$ solution using various inorganic reducing agents.[26] The as-prepared sample in amorphous nature contains 30% of water and a trace amount of oxygen. The sharp weight loss of about 3% occurring at around 525°C in Thermogravimetric analysis corresponded with the loss of oxygen from the lattice to give Mn_2O_3. The SC of nanocrystalline MnO_2 coated on a stainless steel (SS) electrode was found to be 250 F g^{-1} in 0.1 M Na_2SO_4 electrolyte between −0.2 and 1.0 V versus standard calomel electrode (SCE) at a current density of 0.5 mA cm^{-2}.[26] Belanger et al. have reported decomposition of α-MnO_2 into well-crystalline Mn_2O_3 at 400°C.[31] Structural evolution

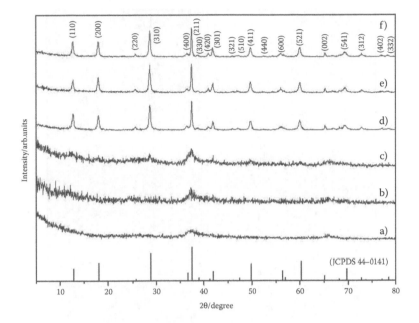

FIGURE 5.8
X-ray diffraction patterns (CuKα = 1.5418 Å) of MnO$_2$ samples as-prepared and dried at 50°C a) in air and annealed at b) 200, c) 300, d) 400, e) 500, and f) 600°C for 3 h in air. The Joint committee for powder diffraction standards (JCPDS) pattern of α-MnO$_2$ is shown at the bottom. The hkl planes are indicated at the top. (From Ragupathy, P. et al., *Journal of the Electrochemical Society*, 155, A34–A40, 2008. With permission.)

from α-MnO$_2$ (as prepared) to Mn$_2$O$_3$ (500–800°C) and then to Mn$_3$O$_4$ (900°C) has been also demonstrated by Devaraj and Munichandraiah.[32] In contrast to these results, a report is available revealing the formation of stable α-MnO$_2$ even at 600°C. The powder x-ray diffraction pattern of the as-prepared and annealed MnO$_2$ is depicted in Figure 5.8. The as-prepared oxide and samples annealed up to 300°C exhibited only one unsymmetrical broad peak around 37.5° (2θ) indicating the amorphous nature of the sample. A gradual evolution from amorphous phase to crystalline phase is observed when the annealing temperature increases to 400°C. All the diffraction peaks are indexed (Figure 5.8) to tetragonal phase (space group I4/m) with the lattice constant of a = 9.8172 Å and c = 2.8582 Å as reported in the JCPDS No. 44–0141.

In addition to this structural evolution, morphology and composition are becoming significantly important to their capacitive performance. SEM images of the as-prepared MnO$_2$ and MnO$_2$ annealed at different temperatures are shown in Figure 5.9. Amorphous MnO$_2$ possess clustered granules of particle size varying between 5 and 10 nm. Gradual growth of nanorods (NR) began when the annealing temperature reached 400–600°C, with well-defined NRs of length 500–750 nm and diameter 50–100 nm, as shown in Figure 5.9.

FIGURE 5.9

SEM images of MnO_2 a) as prepared and dried at 50°C in air (inset shows energy-dispersive x-ray) and annealed at b) 200, c) 300, d) 400, e) 500, and f) 600°C for 3 h in air. (From Ragupathy, P. et al., *Journal of the Electrochemical Society*, 155, A34–A40, 2008. With permission.)

The capacitive behavior of hydrated MnO_2 is more sensitive to its microstructure and water content.[27-33] It is generally recognized that the surface area will enhance the SC of oxides. Brunauer-Emmett-Teller (BET) isotherm exhibited a hysteresis loop around P/P_0 by 0.8–1.0 for the as-prepared MnO_2, indicating the formation of secondary mesopores between the particles.[34] The BET surface area of amorphous MnO_2 is found to be 230 m² g⁻¹ and corresponding pore size distribution with an average pore size of about 14.5 nm. It is presumed that the existence of secondary pores plays an important role, in part, in enhancing the electrochemical activity of the nanostructured MnO_2. However, the lower surface area of annealed samples heated at various temperatures is evident from the absence of the hysteresis loop.

X-ray photoelectron spectroscopy and thermal gravimetric analysis have been employed to establish that the as-prepared MnO_2 contains a significant amount of water molecules and the manganese valence is 4+.

The SC of hydrated MnO_2 was found to be about 250 F g^{-1} in 0.1 M Na_2SO_4 at a current density of 0.5 mA cm^{-2}. It is interesting to note the effect of annealing temperatures of MnO_2 on SC values. Electrodes fabricated with MnO_2 annealed at different temperatures were subjected to charge–discharge studies. The as-prepared sample offered high SCs due to high surface area and substantial water content, which is very essential for transport of active ionic species. As annealing temperature increased to above 400°C, the SC values decreased due to a decrease in the surface area. The sample annealed at 400, 500, and 600°C left no water content, which eventually led to a drastic decrease in SC (Figure 5.10a).

The SC values were recalculated in terms of capacitance density (μF cm^{-2}) using SC and the surface area. The data were plotted as function of annealing temperature as shown in Figure 5.10b. Two different segments are shown in this graph. For MnO_2 samples heated to 300°C, which were predominantly in the amorphous state, capacitance density was about 100 μF cm^{-2}. For the crystalline samples annealed between 400 and 600°C, the capacitance density was about 300 μF cm^{-2}. The increase in capacitance density suggested that Reaction 5.2, corresponding to bulk, governs the crystalline phase and Reaction 5.1 was mainly on the surface in the amorphous phase.

Very interestingly, the amorphous MnO_2 obtained by the reduction of $KMnO_4$ with ethylene glycol exhibited remarkable capacity retention upon cycling.[28] Figure 5.11 shows the cyclic voltammetry (CV) of nano-MnO_2 electrode recorded at 1st, 1200th, and 2000th cycles having a rectangular shape due to the ideal capacitive behavior of its oxides. These three voltammograms almost merge with each other, indicating good stability of the nano-MnO_2 electrode even after it was cycled 2000 times. The capacitance retention upon charge–discharge cycling is strongly attributed to the nature of nano-MnO_2 possessing a high surface area, secondary pores between particles, and appropriate water content. By assumption that a high SC is closely related to the surface area, secondary pores, and water content, which support the proposed surface effect in operating mechanism as an electrode material for supercapacitor, these properties decrease by increasing temperature. It is worthy of note that present-day nano-MnO_2 possesses remarkable stability of SC of about 250 F g^{-1} up to 1200 cycles and a very slight decrease (8%) further up to 2000 cycles (Figure 5.11b).

Recently, hollow micro/nanostructures have been extensively investigated for applications such as batteries, supercapacitors and chemical sensors due to their unique structural features such as high surface area, low density, and permeability. Hollow spheres and urchins of α-MnO_2 have been prepared through simple hydrothermal method without employing any template. The hollow sphere or urchin-structured α-MnO_2 exhibited a highly loose mesoporous structure consisting of thin plates or nanowires. The SCs of hollow α-MnO_2 were 167, 147, and 124 F g^{-1} at 2.5, 5, and 10 mA, indicating the significant rate capability of the materials. Electrode materials for

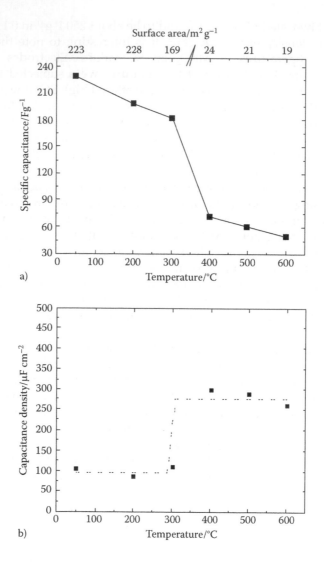

FIGURE 5.10
a) Dependence of SC on temperature and BET surface area at current density of 1 mA cm^{-2}. Annealing temperatures are indicated at each point. b) Capacitance density on annealing temperature at current density of 1 mA cm^{-2}. Potential window: -0.2 to 1.0 V versus SCE; electrolyte 0.1 M Na$_2$SO$_4$. (From Ragupathy, P. et al., *Journal of the Electrochemical Society*, 155, A34–A40, 2008. With permission.)

an excellent supercapacitor should possess not only high capacitance but also low electronic resistance. Electrochemical impedance spectroscopy (EIS) is an experimental method to characterize the electrochemical systems. The similar values of intercept at high frequency for all the samples obtained by hydrothermal process indicate the same combination resistance of ionic

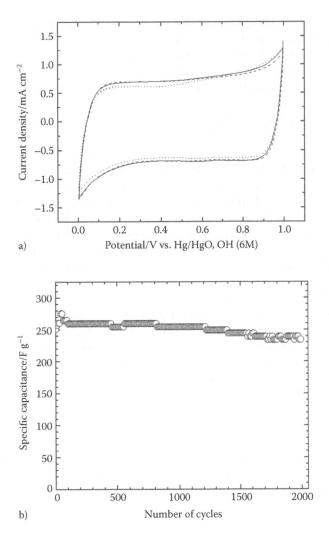

FIGURE 5.11
Cyclic voltammograms in 0.1 M Na_2SO_4 at 10 mV s^{-1} of first (solid line), 1200th (dashed line), and 2000th (solid line) cycle a) and SC as a function of cycle number b) for nano-MnO_2. (From Ragupathy, P. et al., *Journal of Physical Chemistry C*, 113, 6303–6309, 2009. With permission.)

resistance of electrolyte, intrinsic resistance of active materials, and contact resistance at the active material–current collector interface. The semicircle in the high frequency range associated with the faradaic charge–transfer resistance (R_{ct}). The charge–transfer resistance of the α-MnO_2 was in the order of 6 h < 12 h < 3 h < 24 h prepared materials. Moreover, it is inferred from EIS spectra that hollow spheres/urchins possess minimum R_{ct} and diffusive resistance indicating the excellent porous structure of the material.

Munaiah et al. have synthesized nearly x-ray amorphous MnO_2 by a carboxylic acid-mediated system.[35] Various extrinsic and intrinsic properties are responsible for the structural morphologies. It is generally believed that morphology and crystal growth is dependent on the degree of supersaturation, the species around the surface of the crystal, and the interfacial energies. The evolution of the hollow spheres is based on the molecular chemistry of metal oxides in which the emulsion plays a vital role in preferential growth. The reduction of $KMnO_4$ with $Na_2S_2O_4$ in the microemulsion medium produced MnO_2 nuclei, which promote the formation of the large amounts of lamellar nanoplatelets. The formed nanoparticles can aggregate and self-assemble on the surface of the microemulsion. Hollow-sphere manganese oxides were obtained after washing the carboxylic acid with deionized water and were subjected to supercapacitor studies in both univalent cation- and bivalent cation-containing electrolytes. Very interestingly, the variation in voltammetric current density of the electrodes containing bivalent and univalent cations in the electrolyte suggested that the storage capacity depended not only on the microstructure of the materials but also on the nature of the electrolyte. The higher voltammetric current density in cyclic voltammograms and longer discharge time in charge–discharge curves clearly indicated that the storage capacity was higher for the electrodes containing the bivalent cation than for the electrodes containing univalent cation. Thus, the high SC in the bivalent cation-containing electrolyte was attributed mainly to the fact that each bivalent cation reduces two Mn^{4+} to Mn^{3+} by doubling the number of electrons, while univalent cation reduces one Mn^{4+} to Mn^{3+}.[36–38]

On the top of morphology, microstructure, and compositions, the SC depends on the crystallographic form of MnO_2. The capacitance properties are due to intercalation/deintercalation of protons or cations in different crystallographic MnO_2 structures, which possess sufficient gaps to accommodate ions. Devaraj and Munichandraiah intricately studied the effect of crystallographic structures of MnO_2 on its capacitance behavior.[37] With an aim to understand the role of crystallography on capacitor performance, various phases of MnO_2 in nanodimensions were prepared. The existence of various crystallographic structures obtained using different synthetic procedures was confirmed by powder x-ray diffraction patterns. SEM images of α-, β-, γ-, δ-, and λ-MnO_2 are shown in Figure 5.12 (a–e). α-MnO_2 is composed of spherical particles without interparticle boundaries, whereas β-MnO_2 possesses 1D NRs in which the diameter and length were 50 nm and several micrometers as shown in Figure 5.12b. The morphology of γ-MnO_2 is spherical brushes with straight and radially grown NRs differing from α and β microstructures. δ-MnO_2 consisted of aggregated spherical particles made of interlocked short fibers of 10–20 nm in diameter. λ-MnO_2 exhibited polygon-shaped particles with a size ranging between few nanometers and micrometers.

Electrodes fabricated with different crystallographic forms of MnO_2 were subjected to electrochemical studies in aqueous 0.1 M Na_2SO_4 electrolyte.

FIGURE 5.12
SEM micrographs of α a), β b), γ c), δ d) and e) λ-MnO$_2$. (From Xu, C. et al., *Journal of Power Sources*, 196, 7854–7859, 2011. With permission.)

The cyclic voltammograms of various crystallographic MnO$_2$ recorded between 0 and 1 V versus SCE at a sweep rate of 20 mV s^{-1} are shown in Figure 5.13a. The rectangular shape of all the voltammogram showed the ideal capacitive behavior of different crystallographic forms. However, α-MnO$_2$ offered high voltammetric current density compared with other samples due to higher porosity and a greater surface area. An increase in current near 0 and 1 V showed that overpotential for hydrogen evolution reaction as well as the oxygen evolution reaction were lower for γ-MnO$_2$. Quantitative measurement of SC was made from charge–discharge studies. The variation in charge–discharge cycles for different crystallographic forms suggested that the SC values were different as observed in CVs. The SC values of α-MnO$_2$ and δ-MnO$_2$ were found to be 240 and 236 F g^{-1}, respectively, at a current density of 0.5 mA cm^{-2}. Alternatively, only the SC values of 9 and 21 F g^{-1} were obtained for β-MnO$_2$ and λ-MnO$_2$, respectively. It is understood that SC values largely depend on crystal structure rather than surface area while making comparison among various structures. However, within the same structure, surface area and pore size have some

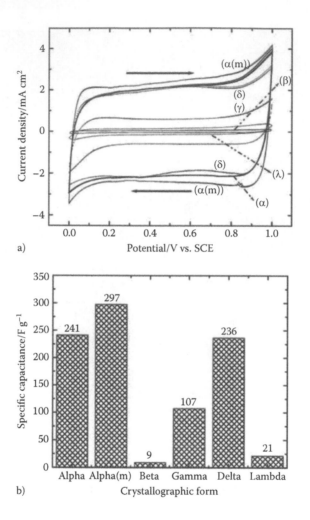

FIGURE 5.13
Cyclic voltammograms of α-, α(m)-, β-, γ-, δ-, and λ-MnO$_2$ recorded between 0 and 1.0 V vs.
SCE in aqueous 0.1 M Na$_2$SO$_4$ at a sweep rate of 20 mVsec^{-1} a) and Specific capacitance of
α-, α(m)-, β-, γ-, δ-, and λ-MnO$_2$ electrodes in 0.1 M Na$_2$SO$_4$ at a current density of 0.5 mA cm^{-2}
between 0 and 1.0 V vs. SCE b). (From Devaraj, S. and N. Munichandraiah, *Journal of Physical
Chemistry C*, 112, 4406–4417, 2008. With permission.)

influence on the variation in the SC value as described elsewhere in the
chapter. For instance, the SC values of α-MnO$_2$ (m) obtained by microemulsion route and α-MnO$_2$ from aqueous-based method were estimated to be
297 and 240 F g^{-1}, respectively. This enhancement in capacitance is mainly
attributed to the greater surface area and higher porosity of α-MnO$_2$ (m)
compared with α-MnO$_2$. From CVs and charge–discharge studies, it is very
clear that SC values of MnO$_2$ decrease in the following order α ∼ δ > γ > λ > β
as shown in Figure 5.13b.

Ye et al. accomplished in preparing the rod-shaped MnO_2 for electrochemical capacitor at different aqueous electrolyte.[40] A maximum SC of 398 F g^{-1} was found in 2 M $(NH_4)_2SO_4$ at 10 mA with excellent cyclability. Nanowires of α-MnO_2 were obtained by acidification of $KMnO_4$ with HCl by Chen et al.[41] MnO_2 nanowires exhibited capacitive behavior in aqueous Na_2SO_4 as electrolyte in the range of 0–0.85 V versus SCE. Wang et al. employed sol-gel template synthesis to prepare highly ordered MnO_2 nanowire arrays as an electrode material for supercapacitor applications.[42] The CV of MnO_2 nanowires with SC 165 F g^{-1} has shown that α-MnO_2 is a promising electrode material for supercapacitors. Subramanian et al. hydrothermally prepared nanostructured MnO_2 for supercapacitor electrodes.[43] The SC for this oxide in aqueous 1 M Na_2SO_4 was found to be 168 F g^{-1}. However, further development in MnO_2-based supercapacitors for commercialization is encumbered due to limited specific capacitance, poor rate capability, and cyclability.

5.4   Nickel Oxide

Nickel oxide (NiO) is of particular interest as pseudocapacitor electrode materials due to its large surface area, high theoretical SC of about 2573 F g^{-1}, high chemical and thermal stability, low cost, and environmental benignity compared with the state-of-art supercapacitor materials like RuO_2.[44–48] Charge storage mechanism of NiO is very different from that of the double-layer capacitors as observed in the high surface area carbon materials. The quasireversible redox process of NiO is responsible for its charge storage property as shown below:

$$NiO + OH^- \leftrightarrow NiOOH + e^- \qquad (5.3)$$

In the past few decades, considerable effort has been taken to develop effective methods to synthesize NiO nanostructures with different shapes and sizes, especially, various novel hierarchical superstructures. Ding et al. reported the synthesis and characterization of NiO nanosheet hollow spheres (NSHS) as electrodes for supercapacitor.[49] The average specific capacities of NSHS were found to be 866, 720, 573, 402, and 259 F g^{-1} at a scan rate of 1, 2, 3, 5, and 10 mV s^{-1} in 2 M KOH electrolyte.

Moreover, the capacity retention at the end of 1000 cycles was found to be 91%, indicating the excellent cyclability. Various NiO hierarchical nanostructures such as nanowires/nanotubes/NRs have been prepared by facile hydrothermal methods followed by thermal treatment.[50]

Pseudocapacitance measurement of NiO nanostructure clearly demonstrated that the nanotubes of NiO exhibited an SC of about 405 F g^{-1} at a current density of 0.5 A/g with good cyclability. The enhanced performance of nanotube NiO is mainly ascribed to their small diffusion path lengths for electrolyte and ensured enough electrolyte contact layer surface of electroactive nanotube arrays. Wang et al. prepared a NiO nanobelt using $NiSO_4.6H_2O$, glycerol, and urea.[51] SEM images show that 1D belt-like morphology can be retained after thermal treatment if no apparent collapse is observed. These nanobelts were utilized as active material for supercapacitor applications. NiO nanobelt offered a high SC of about 600 F g^{-1} at a very high current density of 5 A g^{-1}. It exhibits remarkable capacity retention of about 95% after 2000 cycles. Very high surface area and porous structure of the sample granting complete accessibility by the electrolyte to the active material are evident for its enhancement pseudocapacitive performance. Recently, 1D hierarchical hollow nanostructures have been synthesized using NiO nanosheets on carbon nano-fiber.[52] Due to the advantages of high electroactive surface area, interconnected 2D sheet-like sub-units deliver a high SC of about 702 F g^{-1} at a current density 3 A g^{-1}. Moreover, the hierarchical structure with a hollow interior could increase the amount of electrolyte being accessed throughout the sample.

5.5 Cobalt Oxides

Favorable pseudocapacitance value, good redox, and easily tunable surface of Co_3O_4 have been appealing for supercapacitor applications. The surface faradaic redox process of Co_3O_4 can be described as follows:[53–55]

$$CoO_4 + OH^- + H_2O \leftrightarrow CoOOH + e^- \tag{5.4}$$

$$CoOOH + OH^-O \leftrightarrow CoO_2H_2O + e^- \tag{5.5}$$

In general, by minimizing the dimension of electroactive Co_3O_4 to nanoscale, efficient energy storage can be achieved in supercapacitor applications. For instance, Meher et al. demonstrated the synthesis of tunable dimensionality of Co_3O_4 nanowires by reflux and microwave method.[56] The SC of Co_3O_4 prepared by reflux method was estimated to be 336, 328, 300, 268, and 227 F/g at current densities of 1, 2, 4, 8, and 16 A/g, respectively, while Co_3O_4 obtained by microwave condition offers 232, 227, 200, 168, and 125 F/g at similar current densities. This remarkable enhancement in the SC was attributed to low-dimensional nanowires, which facilitate the effective utilization of electrode active surfaces, low mass transfer resistance, better electrolyte penetration, and faster ion diffusion. Supercapacitor performance of mesoporous Co_3O_4 obtained by water-controlled precipitate methods revealed high SC. This was attributed to the higher surface area, low faradaic charge transfer resistance and enhanced utilization of active material. This helped to sustain enough OH^- supply to ensure the maximum charge storage property.[57]

Yuan et al. developed a facile, cost-effective, and scalable route to prepare well-designed ultrathin Co_3O_4 nanosheet arrays on Ni foam.[58,59] Three-dimensional

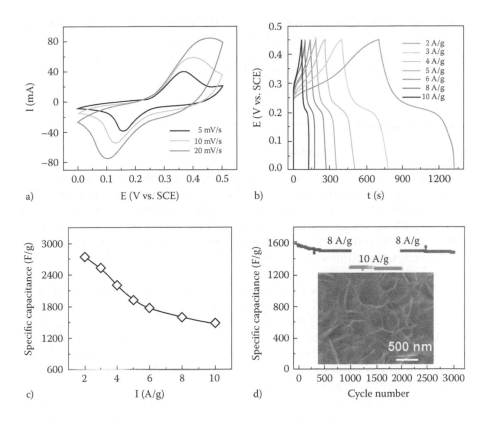

FIGURE 5.14
CV curves a), charge–discharge curves b), SC versus current densities c) and cycling performance d) of the Co_3O_4 nanosheet arrays/Ni foam electrode at varying current densities. The inset in d) shows a top view FESEM image of the Co_3O_4 nanosheet arrays/Ni foam electrode after cycling. (Adapted from Yuan, C. et al., *Energy & Environmental Science* 5, 7883–7887, 2012.)

hierarchical Co_3O_4 nanosheets were grown uniformly on the skeletons of the aslant or perpendicular to the substrate and interconnected with each other to form an ordered nanoarray with highly open porous structures.

Pseudofaradaic reactions were clearly observed in the CV, and charge–discharge data in the potential range of 0–0.5 V versus SCE as shown in Figure 5.14a. The SC of the as-prepared ultrathin porous Co_3O_4 nanosheet was found to be 2735 F g^{-1} at a current density of 2 A g^{-1}. In order to understand the rate capability of Co_3O_4 nanosheets, electrodes made with Co_3O_4 nanosheets were subjected to charge–discharge studies at different current densities (Figure 5.14b and c). It was shown that a very high SC can be still maintained even at a high current density of 10 F/g, indicating the extraordinary rate performance of the nanosheets of Co_3O_4. The remarkable electrochemical performance of Co_3O_4 nanostructures in terms of very high SC good rate capability, and cyclability can be ascribed to its unique 3D hierarchical structure on Ni foam as seen in Figure 5.14d. Particularly, Ni foam with microscale voids and zigzag flow channels ensures the effective mass transport and large surface area per unit area. Moreover, 3D interconnected mesoporous nature of oxide nanosheets increases electrode–electrolyte contact area and ion diffusion for fast electrochemical kinetics.

5.6 Molybdenum Oxides

Molybdenum oxides are appealing as supercapacitor materials due to their low cost and good electronic conductivity arising from strong metallic Mo–Mo bonds. MoO_2 is suitable for electrochemical energy storage systems without adding any conductive carbon such as carbon blocks, carbon nanotubes, and carbon fibers.[60] The pseudocapacitance behavior of MoO_2 is based on the redox transitions $Mo^{4+} \rightarrow Mo^{6+}$.[61] However, the reports available on nanostructured MoO_x for supercapacitors are scarce.

1D NRs of MoO_2 obtained by thermal decomposition of tetrabutylammonium hexamolybdate in N_2 atmosphere offered an SC of 140 F g^{-1} at a current density of 1 mA cm^{-2}.[62] Zhen et al. have succeeded in aligning MoO_3 and MoO_2 NRs on Cu substrate for supercapacitors. The SC values of MoO_3 and MoO_2 were found to be 127 and 205 F g^{-1}, respectively, at constant current 1 mA. These values are obviously higher than the previously reported values.[62] Recently, highly ordered mesoporous MoO_2 has been prepared by silica KIT-16 as hard template. Charge storage mechanism of MoO_2 was investigated using an electrochemical quartz crystal microbalance. The SC value obtained for mesoporous MoO_2 was found to be 200 F g^{-1} at a current density of 1 A g^{-1}. Moreover, this mesoporous nature offers excellent cyclability with no capacity loss even after 1000 cycles.

5.7 Vanadium Pentoxide

Vanadium oxides have also been considered as potential candidate for supercapacitors due to their variable oxidation states, wide potential window, and chemical stability. The first report on vanadium pentoxide (V_2O_5) as capacitor electrode was documented by Lee and Goodenough.[63] The SC value was found to be 350 F/g in aqueous KCl solution. Nanostructured V_2O_5 is always considered an attractive material for supercapacitors due to its high surface-to-volume ratio and maximum electrolyte wet ability. For instance, Wee et al. have reported nanofibers of V_2O_5 by electrospinning technique.[64] Electrospun V_2O_5 yields a reasonably high SC and energy density about 190 F g^{-1} and 5 Wh kg^{-1}, respectively, in 2 M KCl electrolyte. These values can be further enhanced to 250 F g^{-1} and 78 Wh kg^{-1} in Li-ion-containing organic electrolyte.

Hollow sphere V_2O_5 were obtained by a solvothermal method using ammonium metavanadate and ethylene glycol.[65] V_2O_5 hollow spheres exhibit a high SC of 479 F g^{-1} at 5 mV s^{-1} in 5 M LiNO$_3$. This value can be increased to 559 F g^{-1} at a current density of 3 A g^{-1} by modifying the surface of hollow spheres of V_2O_5. This enhanced performance is mainly attributed to the effect of polypyrrole coating, which significantly reduces the phase transformation. Saravankumar et al. have developed supercapacitors utilizing interconnected nanoporous V_2O_5.[66–70] This nanoporous network creates easy ion diffusion into the active materials, resulting in superior performance in terms of capacitance (316 F g^{-1}) and energy density (43.8 Wh kg^{-1}).

5.8   Tin Oxide

Tin oxide (SnO_2), although not widely used, is considered an alternative to the other known supercapacitor materials. In literature, reports of electrochemically and sol-gel prepared SnO_2 in supercapacitors can be found.[71–75] Tin oxides have been electrodeposited on SS electrodes. They showed a rough, highly porous, and nanostructured morphology with many small nanowires, with an SC of 285 F g^{-1} at a scan rate of 10 mVs^{-1} in 0.1 M Na$_2$SO$_4$. At a higher scan rate of 200 mV s^{-1}, the same system showed an SC of 101 F g^{-1}, indicating high-power characteristics of the material. The capacitance was found to increase with an increase in the specific mass of SnO_2. The high value of supercapacitance is considered to be due to the comparatively high conductivity of SnO_2 and the nanostructured and microporous morphology. A decrease of 2.3% of capacitance value was observed between 200 and 1000 cycles.

An SC value of 16 F g^{-1} was obtained for Sb-doped SnO_2 nanocrystalline thin film synthesized by sol-gel process.[74] A composite supercapacitor containing sol-gel SnO_2 and electroplated RuO_2 was reported by Kuo and Wu.[75] The composite electrode yielded an SC of 930 F g^{-1}. It exhibited an operating voltage of 1.0 V and good cycling stability. Wu et al.[75] cathodically deposited tin oxide (SnO_x) onto graphite electrode. The deposit showed rough, porous, nanostructured morphology with many tiny nanowires. For the SnO_x coating deposited at 0.2 C cm^2, maximum SC of 298 and 125 F g^{-1} were achieved from CVs at a scan rate of 10, and 200 mV s^{-1} in 0.5 M KCl, respectively. The long cycle life and stability of the SnO_x coatings on graphite were demonstrated. Alkali metal cations play an important role in the charge–discharge electrochemistry of SnO_x coating. A mechanism based on the surface adsorption of electrolyte cations (C^+) such as K^+ on SnO_x could be suggested as follows:

$$(SnO_x)_{surface} + C^+ + e^- \rightarrow (SnO_xC^+)_{surface} \tag{5.6}$$

and simultaneously, the interaction of H⁺ or alkali metal cations (C⁺) in the bulk of the coating upon reduction is followed by deintercalation upon oxidation.

$$SnO_x + H^+ + e^- \rightarrow SnO_xH \tag{5.7}$$

$$SnO_x + C^+ + e^- \rightarrow SnO_xC \tag{5.8}$$

Reasonably high conductivity of SnO_x and the formation of nanostructured and microporous material could be attributed to more electrolyte cations adsorbed on the large surface and the preferred high values of capacitance found in the present study.[75] Similarly Mane et al. electrochemically synthesized hydrophilic and nanocrystalline tin oxide film electrodes on indium doped tin oxide (ITO) substrates at room temperature.[72] An SC of 43.07 F g⁻¹ in 0.1 M NaOH electrolyte was reported.

5.9   Indium Oxide

Indium oxide (In_2O_3) is a transparent conducting oxide, and it finds applications in many fields ranging from transparent electrodes in electroluminescence devices, solar cells,[76–78] antireflection coatings, and optoelectronic devices to gas sensors.[79] However, reports on the synthesis and application of In_2O_3 thin films are not widely found in the literature. Indium oxide has been reported to have uses in solar cell applications as well. The supercapacitors based on In_2O_3 showed performance, varying according to the morphology of the oxide, which in turn was influenced by the synthesis conditions.[80,81] Literature reported the chemical deposition of In_2O_3 on indium tin oxide substrates, which resulted in electrode materials of different morphologies, that is, nanospheres (NSs) and NRs.[76] An increase in the redox reaction active sites in the In_2O_3 electrode of NR morphology, having nanosized pores and voids, resulted in an increase in the inner and outer charges compared with that of the electrodes with NS morphology. The dependence of the performance of the electrodes on the morphology was apparent from the calculation of active sites, indicating an increase in the number of redox reaction sites from 0.02 to 0.03 when the surface morphology changed from that of NSs to NRs. The reported SC values calculated for In_2O_3 electrodes composed of NRs and NSs were 104.9 and 7.6 F g⁻¹, in 1 M Na_2SO_4, respectively, at a constant discharge current density of 8 A g⁻¹. Another study by Prasad et al. reported electrode position of NRs of In_2O_3 of average length 250 nm and an average diameter 50 nm.[77]

The In_2O_3 electrodes in 1 M Na_2SO_3 electrolyte showed an SC of 190 F g^{-1} at a scan rate of 10 mV s^{-1}. The attraction lay in the fact that the electrodes showed a minimal decrease in the SC value even after 1000 cycles, pointing toward high stability.

5.10   Bismuth Oxide

Bismuth oxide (Bi_2O_3) is a well-known transition metal oxide, which has been intensively studied due to its unique thermal and electrical transport properties.[82,83] Moreover, Bi_2O_3 nanoparticles show good electrochemical stability. It has unique properties such as nontoxic nature, excellent chemical inertness, and biocompatibility. Polycrystalline monoclinic Bi_2O_3 thin films were grown on copper substrates at room temperature by electrode position.[82] Though Bi_2O_3 films were crystalline with a monoclinic crystal structure, they were capable of giving an SC of 98 F g^{-1}. Typical cyclic voltammograms[83] of Bi_2O_3 electrode deposited on copper substrate in 1 M NaOH electrolyte at a scan rate of 100 mV s^{-1} showed that the specific and interfacial capacitances decreased from 98 F g^{-1} and 0.022 F cm^{-2} to 60 F g^{-1} and 0.012 F cm^{-2}, respectively, as the scan rate was increased from 20 to 200 mV s^{-1}. The decrease in capacitance was attributed to the presence of inner active sites that could not sustain the redox transitions completely at higher scan rates.

5.11   Iron Oxide

Iron oxide (Fe_3O_4) is another material used for supercapacitor applications. Aqueous Fe_3O_4 (magnetite) supercapacitors based on powder form and mixed with carbon black have been reported.[84–86] These materials in alkali sulfite and sulfate solutions gave a moderately high value for capacitances. The capacitance of the Fe_3O_4 was found to be sensitive to the anion species but not to either alkaline cations or electrolytes with a pH value of less than 11. The studies indicated a capacitance mechanism, which was different from that of either RuO_2 or MnO_2. Synthesis methods based on electroplating have been reported by Wang et al.[86] The resulted oxide film had a granular morphology and exhibited high porosity. The average crystallite size was 12 nm. The Fe_3O_4 electrode exhibited SC values of 170, 25, and 3 F g^{-1} in aqueous 1.0 M Na_2SO_3, Na_2SO_4, and KOH, respectively. The studies indicated the strong specific adsorption of the anion species. In Na_2SO_3, the capacitance had contributions from both EDLC and pseudocapacitance mechanisms, which involved successive reduction of the specifically adsorbed sulfite anions, from SO_2^{-3} through, for example, S^{-2}, and vice versa. In Na_2SO_4, the current is entirely due to EDLC mechanism. However, KOH showed comparatively low capacitance values, and this was explained to be due to the formation of an insulating layer on the magnetite surface.

5.12   Lesser Used Oxides

Besides the aforementioned materials, use of other nanomaterials of different structures has been reported in the literature. Perovskite bismuth iron oxide nanocrystallite has been reported to have shown an SC value of 81 F g^{-1} by electrode position at room temperature with NaOH as the electrolyte.[87] This value is comparable to ruthenium-based perovskite materials.[88] Another group of materials widely used are the ferrites. Ferrites of the general form MFe_2O_4, where M = Mn, Fe, Co or Ni, have been synthesized mainly by wet chemical methods but the only composite that showed supercapacitive properties was $MnFe_2O_4$. The SC value reported was about 100 Fg^{-1} [89,90] and a power density of 10 KWh^{-1}.[91,92] Nickel ferrites have also been reported to show an SC of 354 F g^{-1} at a scan rate of 5 mV s^{-1} in Na_2SO_3 electrolyte.[93]

Ti-V-W-O/Ti oxide-based supercapacitors and Ti/ $(RhO_x+Co_3O_4)$-based supercapacitors are also reported in the literature prepared by dip coating and thermal decomposition methods, respectively. While the former showed an SC of about 125 F^{-1} g only, the latter showed high values of 300–400 Fg^{-1} due to a combination of the double-layer capacitance with the pseudocapacitance provided by Rh redox-type reactions.[89,92] Table 5.2 summarizes the electrochemical performance of different metal oxides.

The greater energy density can be achieved by fabricating the electrodes with nanostructured materials and also designing a new electrode configuration widely referred to as hybrid supercapacitors (HSCs) or Li-ion supercapacitors. This configuration has one capacitor-type electrode as in the electrochemical double-layer capacitance to realize high power density and battery-type electrodes to ensure high energy density. Thus, newly designed Li-ion supercapacitor is an ideal energy storage device to achieve both high power and energy. The key innovation of using HSC lies by coupling nanostructured materials with the proposed new configuration.[67]

Among the various battery-type electrode materials investigated over the years, lithium metal silicates have been identified as very attractive due to their high capacity, rate capability, and cyclability. For instance, Karthikeyan et al. have demonstrated hybrid Li-ion supercapacitor by constructing lithium metal silicates, Li_2MSiO_4 (M: Mn and Fe), as the negative electrode and activated carbon as the positive electrode in a standard nonaqueous

TABLE 5.2

Summary of the Electrochemical Performance of Different Metal Oxides

S. No.	Metal Oxide	Electrolyte	Capacitance	Operating Voltage Window (V)	References
1.	Ruthenium oxide	H_2SO_4 (aq)	500–1200	–1 to +1	94–109
		KCl (aq)	600–800	0 to +1	
2.	Manganese oxide	(a) NaCl (aq)	100–250	0 to +1	110, 111
		(b) KCl (aq)	100–300	0 to +1	
		(c) Na_2SO_4 (aq)	300–500	0 to +0.9	
3.	Nickel oxide	(a) KOH (aq)	200–278	0 to +0.5	112
		(b) LiCl (aq)	80–100	0 to +0.6	
4.	Cobalt oxide	KOH (aq)	100–200	0 to +0.8	113, 114
5.	Tin oxide	(a) Na_2SO_4 (aq)	100–300	0 to +1	115–116
		(b) KCl (aq)	100–200	0 to +1	
		(c) NaOH (aq)	40–70	0 to +1	
6.	Indium oxide	Na_2SO_4 (aq)	70–100	0 to +1	117
7.	Vanadium oxide	(a) KCl (aq)	200–300	0 to +1	118, 119
		(b) $LiClO_4$ in PC (org)	800–900	–	
8.	Iron oxide	(a) Na_2SO_3 (aq)	100–200	0 to +1	120
		(b) Na_2SO_4 (aq)	10–30	0 to +1	
		(c) KOH (aq)	2–4	0 to +1	

electrolyte such as 1.0 M $LiPF_6$ in EC/DMC electrolyte.[68,69] Very interestingly, energy and power densities are found to be in the range of 40–50 Wh Kg^{-1} and 150–200 W Kg^{-1}, respectively.

Composite cathodes containing polyaniline with $Li(Mn_{1/3}Ni_{1/3}Fe_{1/3})O_2$ and activated carbon have been used as negative and positive electrodes to fabricate Li-ion hybrid capacitor.[70] The maximum observed capacitance of 140 F g^{-1} was reported at a current density of 0.72 A g^{-1} with excellent cyclability. The higher surface area of Polyaniline (PANI) matrix, which allows more electrolyte solution to reach the interior portion of the active materials, is responsible for the enhanced performance.

5.13 Conclusion

The aim of this chapter on nanostructured metal oxides for electrochemical capacitors was to highlight the most recent advancements in supercapacitor technology and a fundamental understanding of the structure–property relationship on energy storage mechanism in supercapacitor applications. The significance of nanostructured metal oxides has been evidenced from the improved performance of supercapacitor devices. Many factors that influence the charge storage process in supercapacitors include high surface area, short ionic diffusion path, morphology, large surface active sites, crystal structure, water content, and the amount of conductive carbon. It is worth pointing out that there has been no standard uniform structure and morphology for metal oxides to achieve the maximum capacitance. However, it is generally accepted that the charge storage occurs on the surface in amorphous materials, whereas the energy storage happens due to bulk intercalation/deintercalation of ions in the crystalline materials. The ultimate goal of the nanostructured materials in supercapacitors is the constitution of 2D and 3D nanoarchitectured cells coupled with the newly designed hybrid supercapacitors. Though the full potential of nanostructured materials has been recognized, innovative manufacturing processes are needed further to the development next-generation breakthrough supercapacitor devices. Selection of counter electrodes, current collectors, substrates, electrolytes, and membrane separators need to be investigated extensively. The challenges remain to material scientists and engineers to understand the transport phenomena of electrons and ions during the electrochemical interface process within active materials.

References

1. Whittingham AG, Zawodzinski T. Introduction: Batteries and fuel cells. *Chem Rev* 2004; 104: 4243–4244.
2. Winter M, Brodd RJ. What are batteries, fuel cells, and supercapacitors. *Chem Rev* 2004; 104: 4245–4270.
3. Thomas B. Reddy. *Linden's Handbook of Batteries*, 4th edn. New York: McGraw Hill, 1994.
4. Scrosati B, Vincent CA. *Modern Batteries: An Introduction to Electrochemical Power Sources.* Arnold, 1997.
5. Conway BE, *Electrochemical Supercapacitors.* Kluwer Academic. New York: Publishers/Plenum Press, 1999.
6. Sarangapani S, Tilak BV, Chen CP. Materials for electrochemical capacitors: theoretical and experimental constraints. *J Electrochem Soc* 1996; 143: 3791–3799.
7. Gamby J, Taberna PL, Simon P, et al. Studies and characterisations of various activated carbons used for carbon/carbon supercapacitors. *J Power Sources* 2001; 101: 109–116.
8. Frackowiak E, Béguin F. Carbon materials for the electrochemical storage of energy in capacitors. *Carbon* 2001; 39: 937–950.
9. Ozoliņš V, Zhou F, Asta M. Ruthenia-based electrochemical supercapacitors: Insights from first-principles calculations. *Acc Chem Res* 2013; 46: 1084–1093.
10. Schäfer H, Schneidereit G, Gerhardt W. *Zeitschrift für anorganische und allgemeine Chemie* 1963; 319: 327–336.
11. Campbell PF, Ortner MH, Anderson CJ. Differential thermal analysis and thermogravimetric analysis of fission product oxides and nitrates to 1500°C. *Anal Chem* 1961; 33: 58–61.

12. Zheng JP, Jow . A new charge storage mechanism for electrochemical capacitors. *J Electrochem Soc* 1995; 142: L6–L8.
13. Zheng JP, Cygan PJ, Jow TR. Hydrous ruthenium oxide as an electrode material for electrochemical capacitors. *J Electrochem Soc* 1995; 142: 2699–2703.
14. McKeown DA, Hagans PL, Carette LPL, et al. Structure of hydrous ruthenium oxides: Implications for charge storage. *J Phys Chem B* 1999; 103: 4825–4832.
15. Hu CC, Chang KH, Lin MC, et al. Design and tailoring of the nanotubular arrayed architecture of hydrous RuO_2 for next generation supercapacitors. *Nano Lett* 2006; 6: 2690–2695.
16. Brumbach MT, Alam TM, Kotula PG, et al. Nanostructured ruthenium oxide electrodes via high-temperature molecular templating for use in electrochemical capacitors. *ACS Appl Mater & Interfaces* 2010; 2: 778–787.
17. Zhang J, Ma J, Zhang LL, et al. Template synthesis of tubular ruthenium oxides for supercapacitor applications. *J Phys Chem C* 201; 114: 13608–13613.
18. Lee HY, Goodenough JB. Supercapacitor behavior with KCl electrolyte. *J Solid State Chem* 1999; 144: 220–223.
19. Thackeray MM. Manganese oxides for lithium batteries. *Prog Solid State Chem* 1997; 25: 1–71.
20. Feng Q, Yanagisawa K, Yamasaki N. Hydrothermal soft chemical process for synthesis of manganese oxides with tunnel structures. *J Porous Mater* 1998; 5: 153–162.
21. Brock SL, Duan N, Tian ZR, et al. A review of porous manganese oxide materials. *Chem Mater* 1998; 10: 2619–2628.
22. de Wolff PM. Interpretation of some γ-MnO_2 diffraction patterns. *Acta Crystallogr* 1959; 12: 341–345.
23. Ma R, Bando Y, Zhang L, et al. Layered MnO_2 nanobelts: Hydrothermal synthesis and electrochemical measurements. *Adv Mater* 2004; 16: 918–922.
24. Hunter JC. Preparation of a new crystal form of manganese dioxide: λ-MnO_2. *J Solid State Chem* 1981; 39: 142–147.
25. Kim H, Popov BN. Synthesis and characterization of MnO_2-based mixed oxides as supercapacitors. *J Electrochem Soc* 2003; 150: D56–D62.
26. Jeong YU, Manthiram A. Nanocrystalline manganese oxides for electrochemical capacitors with neutral electrolytes. *J Electrochem Soc* 2002; 149: A1419–A1422.
27. Ragupathy P, Vasan HN, Munichandraiah N. Synthesis and characterization of nano- MnO_2 for electrochemical supercapacitor studies. *J Electrochem Soc* 2008; 155: A34–A40.
28. Ragupathy P, Park DH, Campet G, et al. Remarkable capacity retention of nanostructured manganese oxide upon cycling as an electrode material for supercapacitor. *J Phys Chem C* 2009; 113: 6303–6309.
29. Yang XH, Wang YG, Xiong HM, et al. Interfacial synthesis of porous MnO_2 and its application in electrochemical capacitor. *Electrochem Acta* 2007; 53: 752–757.
30. Yuan C, Gao B, Su L, et al. Interface synthesis of mesoporous MnO_2 and its electrochemical capacitive behaviors. *J Colloid Interface Sci* 2008; 322: 545–550.
31. Toupin M, Brousse T, Bélanger D. Influence of microstucture on the charge storage properties of chemically synthesized manganese oioxide. *Chem Mater* 2002; 14: 3946–3952.
32. Devaraj S, Munichandraiah N. Electrochemical supercapacitor studies of nanostructured α-MnO_2 synthesized by microemulsion method and the Effect of Annealing. *J Electrochem Soc* 2007; 154: A80–A88.

33. Reddy RN, Reddy RG. Synthesis and electrochemical characterization of amorphous MnO_2 electrochemical capacitor electrode material. *J Power Sources* 2004; 132: 315–320.

34. Xu M, Kong L, Zhou W, et al. Hydrothermal synthesis and pseudocapacitance properties of α-MnO_2 hollow spheres and hollow urchins. *J Phys Chem C* 2007; 111: 19141–19147.

35. Munaiah Y, Sundara Raj BG, Prem Kumar T, et al. Facile synthesis of hollow sphere amorphous MnO_2: the formation mechanism, morphology and effect of a bivalent cation-containing electrolyte on its supercapacitive behavior. *J Mater Chem A* 2013; 1: 4300–4306.

36. Xu C, Du H, Li B, et al. Capacitive behavior and charge storage mechanism of manganese dioxide in aqueous solution containing bivalent cations. *J Electrochem Soc* 2009; 156: A73–A78.

37. Xu C, Wei C, Li B, et al. Charge storage mechanism of manganese dioxide for capacitor application: Effect of the mild electrolytes containing alkaline and alkaline-earth metal cations. *J Power Sources* 2011; 196: 7854–7859.

38. Nayak PK, Munichandraiah N. Reversible insertion of a trivalent cation onto MnO_2 leading to enhanced capacitance. *J Electrocheml Soc* 2011; 156: A585–A591.

39. Devaraj S, Munichandraiah N. Effect of crystallographic structure of MnO_2 on its electrochemical capacitance properties. *J Phys Chem C* 2008; 112: 4406–4417.

40. Ye C, Lin ZM, Hui SZ. Electrochemical and capacitance properties of rod-shaped MnO_2 for supercapacitor. *J Electrochem Soc* 2005; 152: A1272–A1278.

41. Chen X, Li X, Jiang Y, et al. Rational synthesis of α-MnO_2 and γ-Mn_2O_3 nanowires with the electrochemical characterization of α-MnO_2 nanowires for supercapacitor. *Solid State Commun* 2005; 136: 94–96.

42. Wang X, Wang X, Huang W, et al. Sol–gel template synthesis of highly ordered MnO_2 nanowire arrays. *J Power Sources* 2005; 140: 211–215.

43. Subramanian V, Zhu H, Wei B. Nanostructured MnO_2: Hydrothermal synthesis and electrochemical properties as a supercapacitor electrode material. *J Power Sources* 2006; 159: 361–364.

44. Nam KW, Kim KB. A Study of the preparation of NiO x electrode via electrochemical route for supercapacitor applications and their charge storage mechanism. *J Electrochem Soc* 2002; 149: A346–A354.

45. Srinivasan V, Weidner JW. Studies on the capacitance of nickel oxide films: Effect of heating temperature and electrolyte concentration. *J Electrochem Soc* 2000; 147: 880–885.

46. Wang YG, Xia YY. Electrochemical capacitance characterization of NiO with ordered mesoporous structure synthesized by template SBA-15. *Electrochem Acta* 2006; 51: 3223–3227.

47. Yuan C, Zhang X, Su L, et al. Facile synthesis and self-assembly of hierarchical porous NiO nano/micro spherical superstructures for high performance supercapacitors. *J Mater Chem* 2009; 19: 5772–5777.

48. Yuan CZ, Gao B, Zhang XG. Electrochemical capacitance of $NiO/Ru_{0.35}V_{0.65}O_2$ asymmetric electrochemical capacitor. *J Power Sources* 2007; 173: 606–612.

49. Ding S, Zhu T, Chen JS, et al. Controlled synthesis of hierarchical NiO nanosheet hollow spheres with enhanced supercapacitive performance. *J Mater Chem* 2011; 21: 6602–6606.

50. Xiong S, Yuan C, Zhang X, et al. Mesoporous NiO with various hierarchical nanostructures by quasi-nanotubes/nanowires/nanorods self-assembly: controllable preparation and application in supercapacitors. *Cryst Eng Comm* 2011; 13: 626–632.

51. Wang B, Chen JS, Wang Z, et al. Green synthesis of NiO nanobelts with exceptional pseudo-capacitive properties. *Adv Energy Mater* 2012; 2: 1188–1192.

52. Zhang G, Yu L, Hoster HE, et al. Synthesis of one-dimensional hierarchical NiO hollow nanostructures with enhanced supercapacitive performance. *Nanoscale* 2013; 5: 877–881.

53. Zheng MB, Cao J, Liao ST, et al. Preparation of mesoporous Co_3O_4 nanoparticles via solid–liquid route and effects of calcination temperature and textural parameters on their electrochemical capacitive behaviors. *J Phys Chem C* 2009; 113: 3887–3894.

54. Palmas S, Ferrara F, Vacca A, et al. Behavior of cobalt oxide electrodes during oxidative processes in alkaline medium. *Electrochem Acta* 2007; 53: 400–406.

55. Srinivasan V, Weidner JW. Capacitance studies of cobalt oxide films formed via electrochemical precipitation. *J Power Sources* 2002; 108: 15–20.

56. Meher SK, Rao GR. Effect of microwave on the nanowire morphology, optical, magnetic, and pseudocapacitance behavior of Co_3O_4. *J Phys Chem C* 2011; 115: 25543–25556.

57. Wang D, Wang Q, Wang T. Morphology-controllable synthesis of cobalt oxalates and their conversion to mesoporous Co_3O_4 nanostructures for application in supercapacitors. *Inorg Chem* 2011; 50: 6482–6492.

58. Yuan C, Yang L, Hou L, et al. Growth of ultrathin mesoporous Co_3O_4 nanosheet arrays on Ni foam for high-performance electrochemical capacitors. *Energy Environ Sci* 2012; 5: 7883–7887.

59. Eyert V, Höck KH, Horn S. Embedded peierls instability and the electronic structure of MoO_2. *J Phys Condens Matter* 2000; 12: 4923.

60. Shi Y, Guo B, Corr SA, et al. Ordered mesoporous metallic MoO_2 materials with highly reversiblelithium storage capacity. *Nano Lett* 2009; 9: 4215–4220.

61. Zhang H, Wang Y, Fachini ER, et al. Electrochemically codeposited platinum/molybdenum oxide electrode for catalytic oxidation of methanol in acid solution. *Electrochem Solid State Lett* 1999; 2: 437–439.

62. Rajeswari J, Kishore PS, Viswanathan B, et al. One-dimensional MoO_2 nanorods for supercapacitor applications. *Electrochem Commun* 2009; 11: 572–575.

63. Lee HY, Goodenough JB. Ideal supercapacitor behavior of amorphous $V_2O_5 \cdot nH_2O$ in potassium chloride (KCl) aqueous solution. *J Solid State Chem* 1999; 148: 81–84.

64. Wee G, Soh HZ, Cheah YL, et al. Synthesis and electrochemical properties of electrospun V_2O_5 nanofibers as supercapacitor electrodes. *J Mater Chem* 2010; 20: 6720–6725.

65. Yang J, Lan T, Liu J, et al. Supercapacitor electrode of hollow spherical V_2O_5 with a high pseudocapacitance in aqueous solution. *Electrochem Acta* 2013; 105: 489–495.

66. Saravanakumar B, Purushothaman kkKK, Muralidharan G. Interconnected V_2O_5 Nanoporous network for high-performance supercapacitors. *ACS Appl Mater Interfaces* 2012; 4: 4484–4490.

67. Amatucci GG, Badway F, Du Pasquier A, et al. An asymmetric hybrid nonaqueous energy storage cell. *J Electrochem Soc* 2001; 148: A930–A939.

68. Karthikeyan K, Aravindan V, Lee SB, et al. A novel asymmetric hybrid superca-pacitor based on Li_2FeSiO_4 and activated carbon electrodes. *J Alloys Compounds* 2010; 504: 224–227.
69. Karthikeyan K, Aravindan V, Lee SB, et al. Electrochemical performance of carbon-coated lithium manganese silicate for asymmetric hybrid supercapaci-tors. *J Power Sources* 2010: 195: 3761–3764.
70. Karthikeyan K, Amaresh S, Aravindan V, et al. Unveiling organic-inorganic hybrids as a cathode material for high performance lithium-ion capacitors. *J Mater Chem A* 2013; 1: 707–714.
71. Kuo SL, Wu NL. Composite supercapacitor containing tin oxide and electro-plated ruthenium oxide-batteries and energy conversion. *Electrochem Solid State Lett* 2003; 6: A85–A87.
72. Prasad KR, Miura N. Electrochemical synthesis and characterization of nano-structured tin oxide for electrochemical redox supercapacitors. *Electrochem Commun* 2004; 6: 849–852.
73. Wu NL. Nanocrystalline oxide supercapacitors. *Mater Chem Phys* 2002; 75: 6–11.
74. Wu M, Zhang L, Wang D, et al. Cathodic deposition and characterization of tin oxide on graphite for electrochemical supercapacitors. *J Power Sources* 2008; 175: 669–674.
75. Mane RS, Chang J, Ham D. Dye-sensitized solar cell and electrochemical super-capacitor applications of electrochemically deposited hydrophilic and nano crystalline tin oxide film electrodes. *Curr App Phys* 2008; 9: 87–91.
76. Chang J, Lee W, Mane RS, et al. Morphology dependent electrochemical super-capacitor properties of indium oxide: Batteries and energy storage. *Electrochem Solid State Lett*.2008; 11: A9–A11.
77. Prasad KR, Koga K, Miura N. Electrochemical deposition of nanostructured indium oxide: High-performance electrode material for redox supercapacitors. *Chem. Mater* 2004; 16: 1845.
78. Kale SS, Mane RS, Lokhande CD, et al. A comparative photo-electrochemical study of compact In_2O_3/In_2S_3 multilayer thin films. *Mater Sci Eng B* 2006; 133: 222–225.
79. Sberveglieri G, Baratto C, Comini E, et al. Vomiero, Sens. Actuators B, Synthesis and characterization of semiconducting nanowires for gas sensing. *Science Direct* 121(2007); 121: 208–213.
80. Niesen TP, De Guire MR. Review depositon of thin films at low temperature from aqueous solution. *Solid State Ionics* 2002; 151: 61–62.
81. Lee W, Mane RS, Lee SH, et al. Enhanced photocurrent generations in $RuL_2(NCS)2/di$-(3-aminopropyl)-viologen self-assembled on In_2O_3 nanorods. *Electrochem Commun* 2007; 9: 1502–1507.
82. Gujar TP, Shinde VR, Lokhande CD, et al. Electrosynthesis of Bi_2O_3 thin films and their use in electrochemical supercapacitors. *J Power Sources* 2006; 161: 1479–1485.
83. Chen X, Chen S, Huang W, et al. Facile preparation of Bi nanoparticles by novel cathodic dispersion of bulk bismuth electrodes. *Electrochem Acta* 2009; 54: 7370–7373.
84. Wu NL, Wang SY, Han CY, et al. Electrochemical capacitor of magnetite in aqueous electrolytes. *J Power Sources* 2003; 113: 173–179.
85. Brousse T, Bélanger D. A Hybrid Fe_3O_4. MnO_2 capacitor in mild aqueous electrolyte: Batteries, fuel cells and energy conversion. *Electrochem Solid State Lett* 2003; 6: A244.

86. Wang SY, Ho KC, Kuo SL, et al. Investigation on capacitance mechanisms of Fe_3O_4 electrochemical capacitors: batteries, fuel cells and energy conversion. *J Electrochem Soc* 2006; 153: A75.
87. Lokhande CD, Gujar TP, Shinde VR, et al. P^{H}Dependent morphological evolution of β-Bi_2O_3/PANI composite for supercapacitor applications. *Electrochem. Commun* 2007; 9: 1805–1809.
88. Park BO, Lokhande CD, Park HS, et al. Electrodeposited ruthenium oxide thin films for supercapacitor. *Mater Chem Phys* 2004; 86: 239–243.
89. Kuo SL, Wu NL. Electrochemical capacitor of $MnFe_2O_4$ with NaCl electrolyte: Batteries, fuel cells and energy conversion. *Electrochem Solid State Lett.*2005; 8: A495–A499.
90. Kuo SL, Wu NL. Electrochemical characterization of $MnFe_2O_4$/carbon black composite aqueous supercapacitors. *J Power Sources* 2006; 162: 1437–1443.
91. Zuo X, Yang A, et al. Large induced magnetic anisotropy in manganese spinel ferrite films. *Appl Phys Lett* 2005; 15: 2505–2507.
92. Takasu Y, Mizutani S, Kumagai M, et al. Ti- V-W-O/Ti Oxide electrodes as candidates for electrochemical supercapacitors. *Electrochem Solid State Lett* 1999; 2: 1–2.
93. de Souza AR, Arashiro E, Golveia H, et al. Hydrothermal synthesis of CuO-SnO_2 and CuO-SnO_2-Fe_2O_3 mixed oxides and their electrochemical characterization in neutral electrolyte. *Electrochem Acta* 2004; 49: 2015–2023.
94. Hu CC, Chang KH. Cyclic voltammetric depostion of hydrous ruthenium oxide for electrochemical capacitors. *J Electrochem Soc* 1999; 146: 2465–2471.
95. Hu CC, Chang KH. Cyclic Voltammetric deposition of hydrous ruthenium oxide for electrochemical supercapacitors: Effects of chloride precursor transformation. *J Power Sources* 2002; 112: 401–409.
96. Hu CC, Chang KH. Synthesis and characterization of (Ru-Sn)O_2 nanoparticles for supercapacitors. *Electrochim Acta* 2000; 45: 2685–2696.
97. Kuo SL, Wu NL. Composite supercapacitor containing tin oxide and electroplated ruthenium oxide-batteries and energy conversion. *Solid State Lett* 2003; 6: A85–A87.
98. Jang JH, Kato A, Machida K, et al. Supercapacitor performance of hydrous ruthenium oxide electrodes prepared by electrophoretic deposition-batteries, fuel cells and energy conversion. *J Electrochem Soc* 2006; 153: A321–A328.
99. Wang CC, Hu CC. Electrochemical and textural characteristics of (RuSn) $O_x \cdot nH_2O$ for supercapacitors: Effects of composition and annealing: Batteries, fuel cells and energy conversion. *J Electrochem Soc* 2005; 152: A370–A376.
100. Kuo SL, Wu NL. Investigation of pseudocapacitive charge-storage reaction of $MnO_2.nH_2O$ supercapacitors in aqueous electrolytes batteries fuel cells and energy conversion. *J Electrochem Soc* 2006; 153: A1317–A1324.
101. Desai BD, Fernandes JB, KamatDalal VN. Manganese dioxide: A review of a battery chemical. Part II. Solid state and electrochemical properties of manganese dioxides. *J Power Sources* 1985; 16: 1–43.
102. Wu YT, Hu CC. Effects of electrochemical activation and multiwall carbon nanotubes on the capacitive characteristics of thick MnO_2 deposits batteries, Fuel cells and energy conversion. *J Electrochem Soc* 2004; 151: A2060–A2066.
103. Chang JK, Sai WT. Microstructure and pseudocapacitive performance of anodically deposited manganese oxides with various heat treatments, batteries. *J Electrochem Soc* 2005; 152: A2063–A2063.

104. Jeong WL, Su-Il In, Jong DK. Remarkable stability of graphene/Ni-Al layered double hydroxide hybrid composites for electrochemical capacitor electrodes. *J Electrochem Sci Technol* 2013; 4: 19–26.

105. Devaraj S, Munichandraiah N. Electrochemical supercapacitor studies of nanostructured α MnO_2 synthesized microemulsion method and the effect of annealing. *Electrochem Commun* 2005; 5: A373–A277.

106. Hu CC, Tsou TW. Anodic deposition of hydrous ruthenium oxide for supercapacitors. *Electrochem Commun* 2002; 4: 105–108.

107. Broughton JN, Brett MJ. Investigation of thin sputtered Mn films for electrochemical capacitors. *Electrochimca Acta* 2004; 49: 4439–4446.

108. Nagarajan N, Humadi H, Zhitomirsky I. Cathodic electro deposition of MnOx films for electrochemical supercapacitors. *Electrochim Acta* 2006; 51: 3039–3045.

109. Zhitomirsky I, Cheong M, Wei J. The cathodic electrodeposition of manganese oxide films for electrochemical supercapacitors. *J Minerals Metals Mater Soc* 2007; 59: 66–69.

110. Kandalkar SG, Gunjakar JL, Lokhande DC. Preparation of cobalt oxide thin films and its use in supercapacitor application. *Appl Surf Sci* 2008; 254: 5540–5544.

111. Chang J, Park M, Ham D, et al. Liquid-phase synthesized mesoporous electrochemical supercapacitors of nickel hydroxide. *Electrochim Acta* 2008; 53: 5016–5021.

112. He KX, Wu QF, Zhang XG, et al. Electrodeposition of nickel and cobalt mixed oxide/carbon nanotube thin films and their charge storage properties, batteries, fuel cells and energy conversion. *J Electrochem Soc* 2006; 153: A1568–A1574.

113. Kuo SL, Wu NL. Composite supercapacitor containing tin oxide and electroplated ruthenium oxide- batteries and energy conversion. *Electrochem Solid State Lett* 2003; 6: A85–A87.

114. Prasad KR, Miura N. Electrochemical synthesis and characterization of nanostructured tin oxide for electrochemical redox supercapacitors. *Electrochem Commun* 2004; 6: 849–852.

115. Wu M, Zhang L, Wang D, et al. Cathodic deposition and characterization of tin oxide coatings on graphite for electrochemical supercapacitors. *J Power Sources* 2008; 175: 669–674.

116. Mane RS, Chang J, Ham D, et al. Effect of nanosize titanium oxide on electrochemical characteristics of activated carbon electrodes. *Curr Appl Phys* 2008; 9: 87–91.

117. Chang J, Lee W, Mane RS, et al. Morphology dependent electrochemical supercapacitor properties of indium oxide: Batteries and energy storage. *Electrochem Solid State Lett* 2008; 11: A9–A11.

118. Lee HY, Goodenough JB. Electrochemical properties of MnO_2/activated carbon nanotube composite as an electrode material for supercapacitor. *J Solid State Chem* 1999; 148: 81–84.

119. Passerini S, Ressler JJ, Le DB, et al. High rate electrodes of V_2O_5 gel. *Electrochim Acta* 1999; 44: 2209–2217.

120. Wang SY, Ho KC, Kuo SL, et al. Investigation of capacitance mechanisms of Fe_3O_4 electrochemical capactior batteries, fuel cells and energy conversion. *J Electrochem Soc* 2006; 153: A75–A80.

6

Future Scope and Directions of Nanotechnology in Creating Next-Generation Supercapacitors

Meisam Valizadeh Kiamahalleh and S. H. S. Zein

CONTENTS

6.1   Nanotechnology and Its Applications

The primary global research scheme of the 21st century is nanotechnology. Looking forward to the future, nanotechnologies' generalized diffusion will seem to turn them into supplies, generating more space for privileged and superior values of applications such as information technology, nanoenergy, nanobiotechnologies, and nanomaterials.[1–5] In general, nanotechnology is the understanding and controlling of the matters of dimensions of approximately 1–100 nm, in which a unique phenomenon facilitates novel applications.[2] The application domains covered by nanotechnology are discussed in detail in this chapter.

6.1.1   Information and Communication Technology

The information and communication technology (ICT) division has undergone quick development, as nowadays social activities are transformed by novel and varied technologies. Hence, fabrication of smaller transistors using advanced fabrication processes has led to the manufacture of faster computers.[3] Currently, the challenges to continue this miniaturization path exist because as the materials are reduced to nanosize, the change in the characteristics of materials has to be determined and subjected by quantum effects. Miniaturized hardware (sensors, readers, displays, and radio transmitters) and personal secured access to equipment (biometric id) and information (digital id) are some of the applications of nanotechnology in ICT.[6–11]

6.1.2   Biosciences and Life Sciences

The combination of nano- and biotechnologies is an encouraging spot with high expectations where the most important steps can be built for the improvement in medical sciences, health treatment, and body repair. Nanomedicine, targeted drug delivery, and programmed tissue engineering are some advances in this field.[4,12,13]

6.1.3   Materials and Manufacturing

Nanotechnology facilitates high-strength, durable and active materials. Lightweight protective clothes, flexible antiballistic textiles, microsensors for body and brain sensing, wearable and/or flexible displays for visual response, exoskeletons, and robotics are some recent developments of nano-structures and nanocomposites.[5,14,15]

6.1.4 Energy or Power

With an increase in the demands of wearable functionalities and electronics, the need for lightweight wearable electric power is extremely critical. Currently, the developments in the area of lightweight powers include flexible solar cells to recharge batteries, fuel cells, and supercapacitors. Hence, unquestionably, energy storage has remained a main challenge in this century. In response to the requirements of modern society and emerging environmental and ecological concerns, it is now essential that new, inexpensive, and eco-friendly systems of energy conversion and storage are established, leading to a quick development of research in the area of energy storage.[6] It is also noteworthy to determine the opportunities and barriers for developing next-generation electrical energy storage products, such as batteries and supercapacitors, based on nanotechnology.

6.2 Current Commercial Activity

Some of the international companies from different parts of the world currently manufacture electrochemical double-layer capacitors (EDLCs) in a commercial capacity. For instance, NEC and Panasonic in Japan have been producing EDLC materials for more than 30 years. American companies like Epcos, ELNA, AVX, and Cooper manufacture components, whereas companies like Evans and Maxwell produce integrated modules that incorporate voltage balancing circuitry. Kold Ban International designed and marketed a supercapacitor module, which is suited specifically for kick-starting the internal combustion engines in cold weather. Cap-XX in Australia and Korean Ness Capacitor Company offer a range of components for supercapacitors, whereas the Canadian manufacturer, Tavrima, produces a different range of supercapacitor modules. However, the Russian company, ESMA, started selling a wide range of EDLC modules for applications of electric vehicles power quality, and for kick-starting combustion engines.[16] Nowadays, assembling of electrochemical supercapacitors in markets is highly based on the electrode components, such as porous carbon materials with high surface area and noble metal oxide systems.[17] For instance, Matsushita Electric Industrial (Panasonic, Japan) produced gold-based capacitors,[17] and Pinnacle Research (USA) produced supercapacitors especially for high-performance military hardware applications.[17,18] The supercapacitors currently available from the above-mentioned companies, as given on their Web sites, are summarized in Table 6.1. The overall market size of all three areas of batteries, fuel cells, and supercapacitors was estimated to be

TABLE 6.1

Summary of Current EDLCs Available Commercially

Company Name	Country	Device Name	Capacitance Range (F)	Voltage Range (V)	Web site
AVX	USA	BestCap	0.022–0.56	3.5–12	www.avxcorp.com
Cap-XX	Australia	Supercapacitor	0.09–2.8	2.25–4.5	www.cap-xx.com
Cooper	USA	PowerStor	0.47–50	2.3–5	www.powerstore.com
ELNA	USA	DynaCap	0.033–100	2.5–6.3	www.wlna-america.com
ESMA	Russia	Capacitor modules	100–5000	12–52	www.esma-cap.com
Epcos	USA	Ultracapacitor	5–5000	2.3–2.5	www.epcose.com
Evans	USA	Capattery	0.01–1.5	5.5–11	www.evanscap.com
Kold Ban	USA	KAPower	1000	12	www.koldban.com
Maxwell	USA	Bootcap	1.8–2000	2.5	www.maxwell.com
NEC	Japan	Supercapacitor	0.01–6.5	3.5–12	www.nec-tokin.net
Ness	Korea	EDLC	10–3500	3	www.nescap.com
Panasonic	Japan	Gold capacitor	0.1–2000	2.3–5.6	www.macopanasonic.co.jp
Tavrima	Canada	Supercapacitor	0.13–160	14–300	www.tavrima.com

Source: Namisnyk A.M., A survey of electrochemical supercapacitor technology. Faculty of Engineering, University of Technology, Sydney, 2003.

Abbreviations: EDLCs, electrochemical double-layer capacitors; F, farad.

US$350 million by the year 2008 and forecasted to reach US$7700 million in 2012.[19] Markets for nanotechnology-enabled energy production and energy storage were expected to take off from 2012. Robust sales growth opportunities are expected, in particular, for rechargeable batteries (lithium-ion) and supercapacitors based on an early adoption of nanomaterials in the processes and productions of these devices. Estimation of the growth for the supercapacitor market was to be over $600 million by the year 2012, whereas the entire market for battery and supercapacitor storage devices is predicted to increase from US$1.5 billion in 2012 to US$8.3 billion in 2016, which is quite a large growth.[20]

The supercapacitors are commercially utilized as power sources for activators, widely used in long-time constant circuits, or as standby power for random access memory devices and telecommunication devices and so on.[17,21,22] A quantitative comparison of the characteristics and performances among battery, capacitor, and supercapacitor is given in Table 6.2.

TABLE 6.2

A Quantitative Comparison of a Capacitor, Supercapacitor, and Battery

Parameters	Capacitor	Supercapacitor	Battery
Charge time (sec)	10^{-6}–10^{-3}	1–30	0.3–3 hrs
Discharge time (sec)	10^{-6}–10^{-3}	1–30	1–5 hrs
Energy density (Wh/kg)	<0.1	1–10	20–100
Power density (W/kg)	>10,000	1000–2000	50–200
Cycle life	>500,000	>100,000	500–2000
Charge/discharge efficiency	~1.0	0.90–0.95	0.7–0.85

Source: From Nuintek. Comparison of capacitor, supercapacitor and battery. 2006. Available from: http://www.nuin.co.kr.

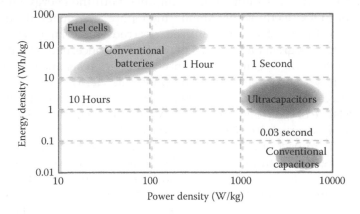

FIGURE 6.1
Ragone plot of different energy storage devices. (From Schneuwly, A., Designing powerful electronic solutions with ultracapacitors. Maxwell Technologies: Rossens, Switzerland, 2006.)

Batteries, in general, are low-power devices in contrast to capacitors, which have power densities as large as 10^4–10^6 W/kg, but lower energy densities. From this perspective, supercapacitors can merge both properties of high power density and higher energy density at the same time. In addition to the high electrical performance, supercapacitors have long life cycles due to the absence of any kind of chemical reactions. Conway carried out a very inclusive and complete review on the historical background, properties, and principles of capacitors.[17] The improvement in the performance of a supercapacitor is presented in Figure 6.1 and the graph is known as "Ragone plot." This kind of graph gives some information about the power densities of a variety of energy storage devices versus their energy densities, measured along the vertical axis and the horizontal axis, respectively. As shown in Figure 6.1, it is believed that supercapacitors occupy an area between conventional capacitors and batteries.[23] Despite having larger capacitances than those of conventional capacitors, supercapacitors are yet to compete with the high–energy–density batteries and fuel cells. Thus, much of the earlier literature investigated for the outline that focuses on developing supercapacitors with the energy level as good as the energy level of batteries.

6.3   Current Research Efforts

A vital aspect in nanotechnology for the supercapacitor applications is to achieve a compromise between two major properties of specific surface area (to guarantee high capacitance) and pore-size distribution (to let electrolyte easily access the electrode surface).[6] Some institutions have been doing researches in order to simultaneously improve both the energy and power densities of EDLC technology. Among high–pore size materials, activated carbons were the most frequently used electrode material in commercial supercapacitors. A great deal of current research is concerned with the issues that contribute to the specific capacitance and series resistance in such electrode materials. Celzard et al. from the Université Henri Poincaré-Nancy in France carried out a research to correlate between the porous structure and series resistance of electrode materials.[25] Another team from the French laboratory at the Conservatoire National de Arts et Métiers performed a research to confirm the impact of pore size distribution on specific capacitance.[26] Xie et al. and Feng and Cummings also believed that electrode materials are the important factors to supercapacitors and an ideal one should have a large surface area and appropriate pore size distribution.[27,28] Xie et al. also stated that from different sorts of good candidate for electrode materials, carbonaceous materials have constantly attracted numerous interests of researchers during the discovery of supercapacitor electrode materials.[29] The early investigations were mainly on activated carbon, mesoporous carbon and carbon nanotubes (CNTs). Recently, more initiatives are taken on understanding the applications of graphene. Yamada et al. compared different carbons of comparable specific surface areas and with different pore size distributions and showed that those that were mostly mesoporous had higher capacitance.[30]

A decade ago, CNTs were discovered and attracted the interests of some academic institutions; for instance, the Poznan University of Technology, Poland, and Sungkyunkwan University in Korea fabricated electrodes from CNTs that exhibit a higher specific capacitance than that attainable by activated carbons.[31,32] Korean researchers at Hanbat National University, Department of Applied Chemistry and Biotechnology, took this a step further and demonstrated that activated CNTs have an even higher specific capacitance than ordinary CNTs.[33] Due to their unique architecture, CNTs are now intensively explored as new electrode material for supercapacitor components, although this brings a high cost issue as for batteries.[6] Some significant interests have also been attracted to conducting polymer materials and researchers recommend that high specific capacitances should be achievable.[34]

Transition metal oxides, after carbonaceous materials, have always been a promising electrode material because of their low resistance and high specific capacitance, but their exorbitant cost has generally hindered their commercial viability options.[18] Conventionally, a strong sulfuric acid was used as an electrolyte with metal oxide electrodes for increasing the ion mobility, which led to an increase in the charge and discharge rates. Nevertheless, this matter brings restrictions for the choice of electrode materials because most of the metal oxide–based electrodes become unstable and corrode in a highly acidic electrolyte. Researchers from the University of Texas, USA, have then paid attention on the possibility of applying a milder, potassium chloride (KCl) aqueous electrolyte in the case of using metal–oxide–based electrodes. Their achievements suggested that the replacement of electrolyte is definitely possible and should extend the accessibility of possible electrode materials further.[35] Manganese oxide (MnO_2), as a more accessible and cheaper substitute to ruthenium oxide, has been proven as a promising electrode candidate by the researchers at the Imperial College, London.[36]

While ceramics with high oxide ion conductivity [especially zirconia (ZrO_2) based] have been recognized for approximately a century, Stotz and Wagner in the 1960s demonstrated that protons do exist as minority charge carriers in oxides.[37] Takahashi and Iwahara later carried out systematic researches on ceramics and showed that acceptor-doped perovskite-type oxides [e.g., doped lanthanum aluminate ($LaAlO_3$), lanthanum yttrium oxide ($LaYO_3$), strontium zirconate ($SrZrO_3$)], which were previously recognized for their moderate oxide ion conductivity,[38,39] could turn into a great proton conductor within the water-containing atmospheres. The experimental conductivities for this ceramic-based electrode were unfortunately still too low in comparison with the large conductivity of yttria (Y_2O_3)-stabilized ZrO_2, within the same standard electrolyte material. However, the related compounds [in particular, acceptor-doped strontium cerate ($SrCeO_3$)[40] and barium cerate ($BaCeO_3$)[41]] with larger proton conductivities were soon found and tested in different sorts of electrochemical cells.

The most attractive results seem to be achieved by hybrid configuration design, which consists of carbon materials and conducting polymers or transition metal oxides. Research on supercapacitor materials at the University of Bologna in Italy demonstrated that using activated carbon as a positive electrode and a polymer-based negative electrode can outperform the configurations of only using activated carbon.[42] Studies at the National Cheng Kung University in Taiwan indicated that high specific capacitance could be attained by depositing conducting polymers onto activated carbon.[43] Frackowiak and coworkers at the Poznan University of Technology also confirmed that coating the CNTs with conducting polymers can increase the electrochemical performances and, hence, the specific capacitance of supercapacitors.[44] The study of solid-state supercapacitors, being conducted at the University of Twente in Netherlands, is also of interest, in which Y_2O_3-stabilized ZrO_2 was used as a replacement for a liquid electrolyte.[45]

Lee's group successfully fabricated a polypyrrole (PPy)–CNT composite electrode on a ceramic fabric by chemical vapor deposition and chemical polymerization technique and it showed promising capacitive behavior and high stability.[46] In addition, the combination of CNTs and porous structure of the ceramic fabrics has granted the promising electrode with a high surface area.

6.4   Future Research

Today electrochemical supercapacitors the great choice for the power sources of hybrid vehicles and all sorts of portable electronics, for instance, cellular phones, notebooks, and current generation of tablets. However, despite having excellent commercial success, these kinds of supercapacitors are still open to development and enhancement. Dynamic research is being done on all aspects of supercapacitors, that is, electrode materials (anodes, cathodes), electrolytes, and cell fabrication. The main parameters restricting their broader application are the cost and safety issues. However, developments continue to take place to address all these limitations.

High–surface area activated carbon is the most frequent anode material utilized extensively in commercially fabricated supercapacitors. In theory, the higher the surface area of the activated carbon, the larger the value of the specific capacitance, although it has low electrochemical cycling stability. To overcome this problem, multiwalled CNTs (MWCNTs) having a superior cycling stability but a relatively low specific capacitance, not greater than 80 F/g, has been used. Thus, if an enhancement in energy content is required, a novel, high-capacity electrode material needs to be developed.

However, activated carbons recently rule the market as an inexpensive electrode material, but evolution in the development of MWCNTs, conducting polymers, and metal oxides is ongoing steadily. The utilization of pseudocapacitive (conducting polymers and transition metal oxides) properties to improve the double-layer capacitance (carbon materials) behavior seems to be a common goal among current researchers; this also enhances the chance of developing the forthcoming generation of supercapacitors with high power and high energy densities.

Several research teams have focused on the development of an alternative electrode material for electrochemical supercapacitors. A variety of transition metal oxides have been introduced to MWCNTs and have been shown to be suitable as electrode materials for electrochemical capacitors. Among the metal–oxide–based materials for application in electrochemical supercapacitors, ruthenium oxide (RuO_2) and iridium oxide (IrO_2) have attained much attention.

RuO_2 has a high double-layer and pseudocapacitance properties (can reach up to ~493.9 F/g)[47] and is quite stable in both aqueous acid and alkaline electrolytes. However, the specific capacitance of this supercapacitor sensitively depends on the technique used for preparation. Unfortunately, disadvantages of RuO_2, such as high cost of the raw material and toxicity, have reduced its great application in supercapacitor electrode materials. Therefore, in this century, great efforts have been undertaken to discover novel and cheaper materials. Several metal oxides and hydroxides, such as those of nickel (Ni), cobalt (Co), vanadium (V), and manganese,[48,49] are being studied extensively. Among these inexpensive metal oxides, MnO_2[33] and nickel oxide (NiO)[27] are believed to be the most promising pseudocapacitor electrode materials due to both their specific capacitance and cost-effectiveness issues.

Recently, ceramic oxides (mixed metal oxides) have also attracted attention as electrode materials in supercapacitors. Gibson and Karthikeyan stated in their patents that improved supercapacitors can be developed to store charge by a combination of faradaic and non-faradaic mechanisms. They have invented ceramic materials having nominal (idealized) compositions that correspond to one of the following groups: ABO_3 (e.g., perovskites), A_2BO_4 (including the alternative form AB_2O_4), and fluorites AO_2, where A and B are metals (A and B = La, Sr, Ca, Mn, Fe, Ni, Co, Ga, Ce, Gd, or any other metal).

Generally, a supercapacitor that uses a transition metal oxide and a conductive polymer as electroactive materials is termed as a pseudocapacitor. A combination of pseudocapacitor and carbon-based electrode material is known as hybrid supercapacitor. In the next section, application of supercapacitor electrode materials and their advantages and disadvantages against the hybrid supercapacitor are investigated in detail.

6.5 Scope of Using Transition Metal Oxides against Carbon- and Polymer-Based Supercapacitor Materials

The basic configuration of a supercapacitor consists of current collectors and electrodes impregnated in an organic or aqueous electrolyte. A membrane is inserted between the two electrodes as a separator to insulate them from each other. The assembly of the supercapacitor unit cell is similarly performed as for the traditional capacitors.[23,50,51] The principle of supercapacitors operation relies on their energy storage mechanism; determining how the distribution of the ions takes place when they come from the electrolyte in the vicinity of the electrode surface. Indeed, the electrode materials and their properties such as high surface area and electrical conductivity are the key components to enhance the efficiency of the supercapacitors.[52,53] This section generally covers some reviews about the classification, charge-storing mechanism, as well as comparative studies to evaluate the performance of different electrode materials in supercapacitors. In the end, the advantages and the current challenges on how to enhance the performance of the super-capacitors is discussed in detail.

6.5.1 Taxonomy of Supercapacitors

According to current R&D trends, the taxonomy of supercapacitors can be divided into three main categories: EDLCs, pseudocapacitors, and hybrid capacitors. Their charge-storing mechanisms are non-faradaic, faradaic, and a combination of these two processes, respectively. Faradaic processes involve the oxidation–reduction reactions in which the charge is transferred between an electrode and electrolyte. In contrast, non-faradaic processes do not contain any chemical mechanism. Hence, the charge distribution takes place on the electrode surfaces by physical processes in which neither formation nor break-ing of chemical bonds is involved. This section will outline the contribution of electrode materials in the performances for each one of these three classes of supercapacitors as well as their subclasses. The different classes and sub-classes of supercapacitors (the taxonomy) are graphically shown in Figure 6.2.

Several works have shown major developments in supercapacitor researches.[50,54–57] At the same time, the disadvantages of supercapacitors including low energy density and high production cost have been identi-fied as major challenges for the furtherance of supercapacitor technolo-gies. To overcome the drawbacks of low energy density, one of the most intensive approaches is the development of new materials for super-capacitor electrodes. To date, carbon materials are the most popular, which have high surface areas for charge storage. In spite of having large spe-cific surface areas, carbon materials are unfortunately limited in charge storage. Supercapacitors based on carbon materials are called electrostatic

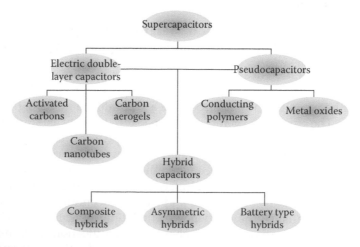

FIGURE 6.2
Taxonomy of supercapacitors. (From Kiamahalleh, M.V. et al., *Nano*. 7, 29, 2012.)

or electric double-layer supercapacitors (EDLS) which have a limited specific capacitance and a low energy density.[57] Hybridizing the electrode materials by adding electrochemically active (pseudocapacitive) materials or completely replacing the carbon materials with electrochemically active materials are the new approaches to enhance the supercapacitors' energy density. Supercapacitors with pseudocapacitive materials as electrodes are called faradaic supercapacitors. Hence, it has been demonstrated that hybrid double-layer supercapacitors can yield much higher specific capacitance and energy density than EDLS.[57]

It is also noteworthy to mention that hybrid supercapacitor with an asymmetrical electrode configuration (e.g., one electrode consists of carbon material while the other consists of pseudocapacitive material) has been extensively investigated recently to enhance overall cell voltage, energy, and power densities.[50,58,59] In this type of hybrid supercapacitor, both EDLC and faradaic capacitance mechanisms take place simultaneously, but one of them plays a greater role.[57]

6.5.2 Electrode Materials

In general, the electrode materials of a supercapacitor can be categorized into three types: (1) carbon materials with large specific surface area,[60] (2) conducting polymers,[61] and (3) metal oxides. In regard to the chemical composition, many kinds of supercapacitor electrode materials have been thoroughly studied, which include electrically conducting metal oxides (e.g., RuO_2,[62,63] IrO_2,[64,65] Fe_2O_3,[66,67] Fe_3O_4,[68] MnO_2,[69–71] NiO,[72–74]), conducting polymers (e.g., polythiophene,[61,75] PPy,[46,76] PANi,[77,78]) and their derivatives, and various kinds of carbonaceous materials (e.g., carbon aerogel,[79,80] activated carbon,[81,82] and CNT).[83–86]

6.5.2.1 Carbon-Based Supercapacitors

Carbon is the most common and economical material for supercapacitor electrodes. Different carbon material–based electrodes have also been intensively investigated.[82,87] Carbon materials, for example, activated carbon, carbon cloth, carbon aerogels, carbon nanofibers, graphite, nanocarbons, and CNTs, generally have larger surface area, in the range of 1000–2000 m^2/g. The capacity of the basal plane and edge plane of graphite carbon is about 10–40 $\mu F/cm^2$ and 50–70 $\mu F/cm^2$, respectively.[88,89] High surface area and porosities can be achieved by carbonization, physical or chemical activation,[90,91] phase separation,[92] gelation,[92] emulsification,[60,93] aerogel–xerogel formation,[94] replication,[95] or burning carbon composites with controllable sizes and volume fractions of open pores, which include adding raw polymer particles, such as poly(methyl methacrylate) spheres,[13] silica sol, and silica gel as templates.[96–98] Generally, carbon-based materials are in powder form and some fabrication techniques are essential to transform these materials into solid packed electrodes. These methods include applying permanent pressure; adding binders such as polytetrafluoroethylene, poly (vinylidene fluoride-hexafluoropropylene), and methylcellulose; and aqueous dispersions of polystyrene, styrene/butadiene copolymer, and ethylene/acrylic acid copolymer, etc.[82,99–104] In theory, the SC value of carbon materials enhances with surface area. However, Qu and Shi[87] and Shi[82] examined many activated carbon materials with various surface areas, pore size, pore size distribution, and pore volume, and correlated these factors with electrochemical capacitance and discovered that the hypothesis is not essentially correct in practical cases.

Frackowiak and Béguin[80] stated that the porous surface of the carbon verifies the ionic conductivity that is related to the ion mobility within the pores. They found that the rate of electrochemical accessibility is mainly indicated by the ion mobility within the pores, which is different from that in bulk electrolytes. Consequently, a resistor network was recommended to be used in the equivalent circuit model instead of a single resistor due to the uneven resistance all through the entire electrode materials. Carbon aerogels are kind of carbon nanoparticles with a massive three-dimensional mesoporous network. These are obtained from the pyrolysis of organic aerogels based on resorcinol-formaldehyde or phenol-furthural precursors using a sol-gel technique. But carbon xerogels[105,106] are produced from a precursor made by conventional drying method, and not using a supercritical method in CO_2. Both carbon aerogel and xerogel are considered as promising electrode materials for supercapacitors due to their large surface area, low density, excellent electrical conductivity, and also for not requiring any additional binding materials. In order to enhance the specific capacitance of carbonaceous materials, functional groups are introduced into the carbon materials because the functional groups are attributed to the pseudocapacitance, which is an extremely efficient method of improving the capacitance behavior. Chu and Kinoshita[107] investigated the effects of pretreatment and surface modification on the electrical double-layer capacitance of a variety of carbonaceous electrodes, and the surface modification of carbon electrodes compromise electrochemical and chemical techniques such as oxidation and reduction and preadsorption of different functional groups at the surface.

6.5.2.2 Conducting Polymer-Based Supercapacitors

Conducting polymers are the third class of candidate materials for supercapacitors because of their excellent electrical conductivity, large pseudocapacitance behavior, and relatively low cost.[108] Polymeric materials, such as p- and n-dopable poly(3-arylthiopene), p-doped PPy, polyacetylene, poly[bis(phenylamino)disulfide], poly(3-methylthiophene), and poly(1,5-diaminoanthraquinone) have been recommended by number of researchers[109–114] as electrodes for electrochemical capacitors.

The most frequently used conducting polymers include PANi,[115] PPy,[116] and poly[3,4-ethylenedioxythiophene] (PEDOT).[117] Cyclic voltammetry (CV) and charge discharge (CD) techniques have been employed to investigate the electrochemical capacitance and charge storage properties of conducting polymers. The typical capacitive behavior of a polymer electrode, however, in general, is not similar to that of an MWCNT electrode, as is expected for a typical capacitor, but exhibits a higher redox potential and pseudocapacitive behavior than an MWCNT electrode. In order to use the same electrode materials on both capacitor electrodes, polymers with a cathodic and an anodic redox process were utilized recently.[118]

Conducting polymers, such as PANi, PPy, polythiophenes, polyacetylene, and poly[bis(phenylamino)disulfide],[119] comprise a large degree of π-orbital conjugations that lead to electronic conductivity and can be oxidized or reduced electrochemically by withdrawal or injection of electrons, respectively. Prasad and Munichandraiah[120] coated the PANi on the stainless steel by a potentiodynamic method from an acidic electrolyte and obtained a very high SC value, up to 450 F/g.

Conducting polymers have a very large SC value that is close to RuO_2, e.g., 775 F/g for PANi,[121] 480 F/g for PPy,[111] and 210 F/g for PEDOT.[122] However, conducting polymers commonly have poor mechanical stability due to repeated intercalation and depletion of ions during charging and discharging.

6.5.2.3 Transition Metal Oxide-Based Supercapacitors

Besides the carbon material electrodes, transition metal oxide electrodes are very important in electrical storage devices. In general, transition metal oxides can provide higher energy density for supercapacitors than conventional carbon materials and better electrochemical stability than polymer materials.[57] They not only store energy like electrostatic carbon materials but also exhibit electrochemical faradaic reactions between electrode materials and ions within appropriate potential windows.[123]

Conducting metal oxides such as RuO_2 or IrO_2 were the favored electrode materials in early electrochemical capacitors used for space or military applications.[124] The high SC value in combination with low resistance resulted in very high specific powers. These capacitors, however, turned out to be too expensive. A rough calculation of the capacitor cost showed that 90% of the cost resided in the electrode material. In addition, these capacitor materials are only suitable for aqueous electrolytes, thus limiting the nominal cell voltage to 1 V.[125,126]

Thus, several attempts were undertaken to obtain the advantages of metal oxides at a reduced cost for a supercapacitor such as Fe_2O_3,[67,127] Fe_3O_4,[68] SnO_2,[128] CuO,[129,130] MnO_2,[69–71,131,132] and NiO.[72,133,134] The dilutions of the costly noble metals, such as Ru, Sr, Ca and La, by forming perovskites ($SrRuO_3$, $CaRuO_3$, and $Sr0.8La0.2RuO_3$) were reported by Guther et al.[135] Other forms of metal compounds, such as nitrides, were also investigated by Liu et al.[136]

Among all, researchers recently have focused on searching for cheaper materials such as MnO_2,[137,138] NiO,[139,140] Co_3O_4,[141,142] and CuO[143] to replace

RuO$_2$, but the selection has traditionally been limited by the use of concentrated sulfuric acid as an electrolyte.[18,144,145] It was believed that high capacitance and fast charging are largely the result of H sorption, so a strong acid was therefore necessary to provide good proton conductivity. However, this resulted in a narrow range of possible electrode materials, as most metal oxides break down quickly in acidic solutions.

Therefore, milder aqueous solutions such as KCl[36] and KOH[33] have been considered for use with metal oxides such as MnO$_2$.

6.5.2.4 Hybrid Nanocomposite-Based Supercapacitors

As discussed, different electrode materials have different strong points and drawbacks. Compared with the conventional capacitors, a higher energy density and a corresponding cycle life can be achieved by hybrid electrochemical supercapacitors that use two different electrode materials.[146,147]

In order to take full advantage of different electrode materials, hybrid nanocomposite supercapacitor materials have been proposed by several researchers. As shown in Figure 6.2, hybrid nanocomposites can be made from different double-layer capacitor and pseudocapacitor materials such as conducting polymer/CNT,[108,148–152] activated carbon/conducting polymer,[43,109,153–155] activated carbon/CNT,[156–160] metal oxide/activated carbon,[161–164] metal oxide/CNT,[103,165–167] mixed metal oxide/CNTs,[168-170] and metal oxide/CNT/conducting polymer.[171–176]

While the electrochemical characteristics of the composite materials are compared with the single element electrode material, one can understand the advantages of such hybrid electrode materials. For instance, Lota and coworkers have shown that the composite of active carbon and NiO exhibits a much more rectangular shape of CV curves compared with that of bare NiO.[177] The shape of a CV curve is a good means of characterizing of a double-layer capacitive behavior. It was also declared that the capacity of the pristine NiO is low and noticeably reduces with increasing current regimes. The bare NiO has only pseudocapacitive properties. In the work by Lota et al., the composite with only 7 wt% of NiO showed the highest value of capacity. They believed that this quantity of NiO was optimal to take advantage of both double-layer capacitive and pseudocapacitive properties of the composite. The active carbon played a critical role as an excellent conducting support for NiO, which also showed quite a high resistivity.

6.6 Advantages of Current Supercapacitors

6.6.1 High Power Density

Supercapacitors display a much higher power delivery (1–10 kW/kg) when compared with lithium ion batteries (150 W/kg). Since a supercapacitor stores electrical charges both at the electrode surface and in the bulk near the surface of the solid electrode, rather than within the entire electrode, the charge–discharge reaction will not necessarily be limited by ionic conduction into the electrode bulk. So, the charging and discharging rates are much faster than the electrochemical redox reactions inside batteries. These rapid rates lead to high power density in supercapacitors.[57] For example, a supercapacitor can be fully charged or discharged in a time that varies from few seconds to several minutes, and the energy can be taken from it very rapidly, say within 0.1 s.[178,179] However, the charging time for batteries is normally on the scale of hours.

Pasquier and coworkers achieved a high specific power of 800 W/kg from a nanostructured $Li_4Ti_5O1_2$ anode and an activated carbon cathode used in a nonaqueous asymmetric hybrid electrochemical supercapacitor.[180] However, prior to this work they used activated carbon[181] and nanostructured lithium titanate oxide[182] as the anode with a poly(fluoro)phenylthiophene as the cathode to investigate the benefit of conducting polymer on the power density. But their polymer-based system had two major drawbacks: a smaller cycle-life and a higher cost than activated carbon. In 2004, they coupled a nanostructured $Li_4Ti_5O1_2$ anode with a poly(methyl)thiophene (PMeT) cathode. The PMeT cathode electrochemically synthesized by the cheap monomer produced a lower cost, higher purity polymer and enhanced the power density to 1000 W/kg.[182]

6.6.2 Long Life Expectancy

In contrast to batteries, supercapacitors have no or negligibly small chemical charge transfer reactions and phase changes in charging and discharging processes so that a supercapacitor can have almost unlimited sustainable cyclability. Moreover, supercapacitors can operate at high rates for 500,000–1,000,000 cycles, with only small changes in their characteristics, but such longevity is impossible in batteries.[57] The life expectancy of a supercapacitor is estimated to be up to 30 years, which is much longer than that of lithium ion batteries, which is only between 5 and 10 years).[57] Although the redox reactions for Faradaic supercapacitors during their charge–recharge process are very quick, their life expectancy is also much longer than that of batteries.[50,56,183]

In a research on the life expectancy for MnO_2-based electrode, by Chang et al.,[184] the stability and cycle life of MnO_2 was investigated 600 times (at a rate of 5 mV/s) and there was no significant capacitance fading detected, which showed the excellent cyclic stability of MnO_2. Their experimental results have also introduced ionic liquid 1-ethyl-3-methylimidazolium–dicyanamide as a potential electrolyte for long-life MnO_2 supercapacitors.[184]

It is noteworthy to know that the specific capacitance of a supercapacitor with an electrode material will fade in the long-term cycle performance. Pristine MnO_2 and $CNTs/MnO_2$ hybrid electrodes had undergone life testing for about 500 cycles by Chen and coworkers.[185] A noticeable decrease in the specific capacitances of both MnO_2 and $CNTs/MnO_2$ hybrid electrodes occurred as the cycle number increased, but fading was less observed with $CNTs/MnO_2$ electrode. Hybrid electrodes exhibited better cyclic efficiency than electrodes with only MnO_2 and it was attributed to the porous structure of $CNTs/MnO_2$ electrodes.[185] Hence, hybridizing pseudocapacitive materials with porous carbon structured materials may tend to provide more space for dissolution and precipitation of pseudocapacitive material particles in the processes of repetitive CV reactions.

6.6.3   Long Shelf Life

Another advantage of supercapacitors is their long shelf life. Most recharge-able batteries will degrade and become noticeably useless if left unused for months, which is due to their gradual self-discharge and corrosion.[57] On the contrary, supercapacitors retain their capacitance and thus are capable of being recharged to their original condition; even though self-discharge over a period of time can lead to a lower voltage. It is reported that supercapaci-tor can stay unused for several years, but still remain close to their original condition.[50]

6.6.4   Wide Range of Operating Temperatures

The ability to perform efficiently over a broad range of temperatures is also an advantage of using supercapacitors. For some remote stations that use battery energy storage system and are located in cold climates, an auxil-iary heating system must be provided to maintain the temperature at close to room temperature. Thus, additional cost and energy consumption are required.[18] Supercapacitors can operate effectively at enormously high and low temperatures. The typical operating temperature for supercapacitors var-ies from −40 to 70°C. Militaries can benefit from this behavior, where reliable energy storage is required to run electronic devices under all temperature conditions during war.[57] Using ionic liquid electrolytes such as ethyl-methyl imidazolium-bis(trifluoro-methane-sulfonyl)imide in supercapacitors is a significant advantage due to its high temperature stability, even at 60°C.[185,186]

Balducci and coworkers[187] investigated the effect of different temperatures on an activated carbon/poly(3-methylthiophene) hybrid supercapacitor using ionic liquid 1-buthyl-3-methyl-imidazolium and organic liquid pro-pylene carbonate-tetraethyl ammonium tetrafluoroborate electrolytes. The ionic liquid electrolytes at room temperature exhibited conductivity values significantly lower than that of organic liquid. However, when the tempera-tures increased to even higher than 60°C, particularly for the electric vehicle applications involving fuel cells, the conductivity values were approximately the same for both ionic and organic electrolytes. This is also attributed to the high electrochemical stability of ionic electrolytes at high temperatures.[186]

6.6.5 Environmental Friendliness

Supercapacitors as an alternative to batteries are more environmentally friendly, are lead free, and have no disposal issues at the end of their life. They do not contain hazardous or toxic materials, and their waste materials are easily disposable.[57] Use of porous carbon materials,[153,171,172,188,189] conducting polymers,[153,172,175,190] and transition-metal oxides[171,172,188,191–193] as single or hybrid electrodes and ionic based liquids[153,193] as electrolytes suggests supercapacitor as the most environmental-friendly and promising energy storage system.

6.7   Challenges for Supercapacitors

Although supercapacitors have many advantages over batteries and fuel cells, they also face some challenges at the current stage of technology.

6.7.1   Low Energy Density

The main challenge for supercapacitor applications in the short and medium terms is their low energy density. Supercapacitors suffer from limited energy density (about 5 W h/kg) when compared with batteries (>50 W h/kg).[57] Commercially available supercapacitors can provide energy densities of only 3 to 4 W h/kg. If a large energy capacity is necessary for an application, a larger supercapacitor must be assembled, which increases the cost.

6.7.2   High Cost

The other important challenges for supercapacitor commercialization are the costs of raw materials and manufacturing. The major cost of supercapacitors arises from their electrode materials. Recently, highly porous carbon materials[23,80,194] and RuO_2[195,196] are the most common electrode materials used practically in fabricating commercial supercapacitors. However, carbon materials, in particular those with a high surface area, are currently expensive (US$50–100 per kg),[50] SWCNTs ~$100,000/kg, MWCNTs ~$5 000/kg, activated carbon ~$15/kg,[197] not to mention the cost of a rare metal oxide like RuO_2. Also, the separator and electrolyte can boost the expense. For instance, if supercapacitors use organic electrolytes, their cost is very high and far from negligible.[57]

6.7.3   Low Operating Voltages

Supercapacitors also have lower operating voltages (typically between 1 and 3.5 V per cell) compared to other types of capacitors and batteries in which the voltage is limited by the breakdown potential of the electrolyte.[57] The solution to meet the expected voltage is that supercapacitors must be series connected like batteries.[198] Commercial supercapacitor cells exhibit the potential window from 0 V to approximately 1 V for aqueous electrolytes and from 0 V to 2.5–2.7 V for organic electrolytes. Maximum voltages for hybrid supercapacitor cells depend upon the electrode materials and the electrolytes.[53] Matsumoto and coworkers[199] showed that the Pt-deposited CNT electrode gave 10% higher voltages than Pt-deposited carbon black. However, the amount of platinum deposited on CNT and carbon black was 12% and 29%, respectively. The performance of the Pt/CNT electrodes was higher than that of the Pt/CB electrodes due to the presence of well-dispersed Pt particles on the CNT surfaces and higher electric conductivity of CNTs.[199]

In summary, to develop new materials with optimal performance, there are two major research directions in the exploration of supercapacitor electrode materials: composite materials and nanomaterials. On the one hand, apart from the type of electrode material used for the supercapacitors, combining the advantages of different materials to make composites should be an important approach to optimize each component for increasing the supercapacitor performance.[200] It is worthy to point out that each component in the composites can have a synergistic effect on the performance of a supercapacitor through minimizing particle size, enhancing specific surface area, inducing porosity, preventing particles from agglomerating, facilitating electron and proton conduction, expanding active sites, extending the potential window, and protecting active materials from mechanical degradation, improving cycling stability, and providing extra pseudocapacitance.[57] A high specific capacitance of 1809 F/g has been reported on using mesoporous Co_xNi_{1-x}-layered double hydroxide composites.[201] The key point to the excellent electrochemical capacitance performance of this composite was attributed to the synthesis of a novel mesoporous microstructure that can accommodate electrolyte and increase the density of active sites for enhancing fluid–solid reactions.[201]

On the other hand, development of nanostructured materials, such as nanoaerogels, nanotubes/rods, nanoplates, and nanospheres, can also have significant enhancement in electrochemical performance by possessing large specific surface area. They can provide short transport–diffusion path lengths for ions and electrons, leading to quicker kinetics, further efficient contact between solid surface and electrolyte ions, and more electroactive sites for faradaic energy storage. This results in high charge–discharge capacities even at high current densities.[57] An extremely large specific capacitance of 3200 F/g was reported by Zhang et al.[202] by porous hybrid materials containing nanoflake-like nickel hydroxide and mesoporous carbon. They attributed the overall enhanced electrochemical behavior to the distinctive structure design of the hybrid material in regard to its nanostructure, large specific surface area, and superior electrical conductivity.[202]

As a result, material morphology is intimately related to the specific surface area and the electrolyte ion diffusion in the electrode. Hence, one-dimensional nanostructure materials are suggested to be highly promising for supercapacitor application due to their reduced diffusion paths and larger specific surface areas.[57]

6.8 Current Applications of Supercapacitors

Supercapacitors are operating the path into more and more applications, which requires electrical energy to be stored. These robust devices can be charged and discharged thousands of times and generally outlive a battery. Many supercapacitor manufacturers claim a life span of 10 years or more. It is also noteworthy that the supercapacitors and batteries are not in competition; rather, they are different products having their unique applications.

Supercapacitors in terms of their power densities are divided into two segments: (1) large cans from 100 F to 5000 F used for heavy-duty purposes, for instance, automotive industry and utilities;[203] (2) small devices in the range of 0.1 F–10 F used in general electronics.[204] A practical way to evaluate and compare supercapacitor performance is to employ a Ragone plot, which plots power density versus energy density (as shown in Figure 6.1).

6.8.1 Memory Backup

Semiconductor memory backup is the most common application of a supercapacitor for any electronic equipment that contains CMOS, RAM, or a microprocessor.[203] For several years, lithium batteries have served as backup power for volatile memory and real-time clock, but they are not always an ideal solution due to their relatively short cycle life and limited operating temperature range. However, their end-of-life disposal issues have to be taken into account.

Supercapacitors, rather than operating as energy storage devices, also perform quite well as low-maintenance memory backup in order to bridge short power interruptions. In 1978, the NEC Company marketed the supercapacitor as computer memory backup for the first time.[203] They obtained the licence for commercializing the invention from the Japanese Standard Oil Company of Ohio, which accidently rediscovered the effect of the EDLC in developing fuel cell designs.[205]

Pierre Mars reported the use of supercapacitor as backup power supply for the cache memory used in solid-state drives (SSDs).[206] The SSD type of hard disks are electrically, mechanically software compatible with a conventional hard disk drive. When cache is used in normal hard disks, a noticeable data transfer speed in terms of reading and writing can be achieved.

Pierre's power supply topology using CAP-XX supercapacitors also confirmed the simulation results. This supercapacitor showed to be highly supportive to the SSD when it flushes the cache and elegantly shuts down with no loss of data when the power fails.[206]

A simplified backup system having a two-cell series supercapacitor charger (LTC$_3$226) with a PowerPath controller was designed by Linear Technology.[207] This backup system consists of a charge pump supercapacitor charger with programmable output voltage and automatic cell voltage balancing, a low dropout regulator and a power-fail comparator in order to switch between normal and backup modes. Three different voltage points simplify and control the backup set up in this system: (1) a high trigger point (3.6V) when the comparator encounters power failure,[203] (2) a standby mode point (3.15V), and (iii) a backup mode point (3.10V) when the system initializes to hold up the power in the absence of battery power. The holdup power for the LTC$_3$226 was reported to be for a time period of about 45 seconds.[207]

6.8.2   Electric Vehicles Power Quality

The use of hybrid power system in vehicles eliminates the need of battery electrification, as these systems are able to recover and restore the energy from several times of braking. The most important advantage is their use in reducing global CO_2-emissions, has become a very critical topic nowadays.[208] However, upgrading the vehicles with hybrid power systems could be quite costly (approximately \$9500 per kWh)[209], but their greater lifetime (10–20 years), cycle life (0.5–1 million charge/discharge cycles), and complete discharge without any life degradation are highly considerable in the economic evaluation.[210] Supercapacitors serve as power quality systems for direct current (DC) motor drives, uninterruptible power supply (UPS) systems, and hybrid and electric vehicles. The automotive industries are also interested in this technology and are currently involved in the use of supercapacitors for the hybrid electric vehicles. Supercapacitors are used in many large-scale applications, particularly transportation industry, for instance, hybrid cars, autonomous rail-guided vehicles, support of substation voltage for trolleybuses, tramways, subways (a prototype being tested), buses, and trains.[208,209,211–213] These ultracapacitors due to their high power density can operate as a supplementary power source to batteries in order to provide a burst of acceleration for the vehicles or can be used to take up and store energy during regenerative breaking system. Many of the hybrid electric vehicles coming to the market employ supercapacitors. For instance, Honda developed their market by employing the supercapacitor technology in many of their vehicles.[214]

Smith and Sen[210] also stated that large-scale battery systems can benefit from setting up the supercapacitors in parallel with them in order to compensate for the quick- and short-term interruptions. Employing this technology may decrease the undue stress being put on the batteries by these temporary interruptions. Lohner and Evers[208] employed a double-layer capacitor- (supercapacitor-) based hybrid power train for light rail vehicles and carried out their first practical test on a diesel-electric city bus.

Most of the driving cycles of vehicles such as trams, light-rail vehicles, and city buses are basically made of two stages: acceleration and deceleration. To date, the greater part of the kinetic energy generated by deceleration on diesel engines has not been used, as they are not able to recuperate energy. On the contrary, trams and light-rail vehicles require an additional accelerating vehicle as an energy consumer in their net part. Researchers have come out with a primary energy saving up to 30% if the power train could store the kinetic energy from the vehicle decelerating stage for the subsequent accelerating stage.[215,216]

The energy saving unit[75] includes a buck-boost inverter and a double-layer capacitor battery, which support the prime mover unit (PMU) during vehicle acceleration and deceleration stages. The supercapacitor can store the recovered kinetic energy from regenerative braking in order to use it for the next acceleration phase.[208] The voltage required for the bus traction is in the range of the battery from 350 V to 720 V, which can be provided by 288 supercapacitor cells with 2.5 V, connected in series. The total storable energy of the battery is able to accelerate the vehicle up to 50 km/h (maximum speed of urban driving cycles) without the help of PMU. The supercapacitors installed on the urban buses benefit from several numbers of stops so that the entire braking energy can be recovered in the battery for several times.[208]

6.8.3 Electromechanical Actuators

Electromechanical actuators directly convert electrical energy into mechanical energy, and a high electric power is required for this conversion. Supercapacitors, having higher capacitances in comparison with that of ordinary conventional capacitors are a better option to be used as electromechanical actuators. Artificial muscles in robots, biomimetic flyings, optical fiber switches, microsensors, optical displays, prosthetic devices, sonar projectors, and microscopic pumps are some of the great potential applications of electromechanical actuators.[217-220] However, these applications are restricted to some extent by the maximum allowable operation temperature and the required voltages.

The electromechanical actuators and supercapacitors both consist of two separate electrodes (anode and cathode), which are separated by an ionic conducting insulating material in the electrochemical cell. Both faradaic and non-faradaic materials have served as actuator electrode materials. For instance, over two decades ago, Baughman and coworkers became pioneers by proposing the conducting polymer actuators based on electrochemical dopant intercalation.[221] Since then, many researchers from different laboratories benefited from the idea of using conducting polymers with faradaic properties such as polyaniline,[222] poly(2-acrylamido-2-methylpropanesulfonic acid) (PAMPS),[223] and PPy.[224]

Transition metal oxides are other types of faradaic materials, which have also been used in actuators. A patent by Park's group from Harvard University reported their invention: nanoscale electromechanical actuator based on individual transition metal oxide nanowires.[225] They used different metal oxides of titanium and zirconium in actuator electrodes. Both titanium and zirconium oxides showed high converse-piezoelectric properties, which make them a promising material to fabricate electromechanical actuators in order to position and move nanometer-sized objects. Recently, Mrunal's research group have shown that the piezoelectric and semiconducting properties of ZnO nanoparticles can effectively improve the recorded motional current and lead to high integrated electrical actuation. However, faradaic materials show battery-like properties storing high energy, but they limit the power density, cycle life, and energy conversion efficiencies of the actuators due to their solid-state dopant diffusion and structural changes. Hence, new actuators have been designed and fabricated having carbon electrode materials with high conductivity and non-faradaic properties[226,227] as they do not have any dopant intercalation.[220]

Generally, the capacitance of an electromechanical actuator similar to supercapacitors directly depends on the separation between the charges on the electrode and the counter charges in the electrolyte and this separation is in the scale of a nanometer. Therefore, carbon nanomaterials, particularly CNTs[218–220] and graphene,[228,229] are the promising candidates for electrode materials to store large capacitances as they possess high surface area accessible to the electrolyte. CNT-based electromechanical actuators were also reported by Terrones and coworkers. This actuator can function at low voltages and temperatures up to 350°C, which indicates the high thermal stability of CNTs and industrial application of carbon electrodes.[230]

As earlier discussed, both faradaic and non-faradaic materials have some advantages and drawbacks. In order to benefit from both of them, syntheses of their composites at nanoscale have been proposed by some researchers. Peng et al. made a nanocomposite by incorporating CNTs and polydiacetylene (PDA), a polymer with interesting electrical and optical properties.[231] This nanocomposite could reversibly change color in response to electrical current and mechanical stress, which made it promising for applications in many fields such as sensors and actuators. Recently, Liang and coworkers have stated that CNTs have poor processability and difficulty of forming stand-alone bulk film materials with pure PDA that hinder its practical use in the field of actuation.[228] Hence, they proposed a two-dimensional (2D) single layer of graphene (relatively smooth surface) as an alternative to CNTs, where the PDA could have well-ordered stripes on its surface. Combining the excellent intrinsic features of graphene with the properties of unique environmental perturbation-actuated deformational isomerization of PDA caused the high conductivity for graphene–PDA nanocomposite and thermal-induced expansion of PDA. The actuation test showed that the actuator movement while applying AC (alternate current) of high frequency AC was quite faster than that at low frequency.[228]

The most important parameter for actuators is the ratio of actuator strain to unit volume of charge in electrode materials. It is also noteworthy that every ion stored in the electrodes creates strain, and the larger the strain is, the better the actuator is as an electromechanical energy conversion device. Liu and coworkers have shown that a relatively large strain can be attained from RuO_2 when it is mixed with carbon material. The high surface area of the metal oxide combined with the high conductivity of carbon material assists in achieving rapid charging and fast motion of the transducer.[232] The ion size is also a parameter that proportionally influences the electromechanical actuator efficiency. Larger the ion size, lesser the mobility and greater the ratio of volume strain/ion in the electrode materials, which will directly lead to higher actuator efficiency.[233] Ionic liquids, as the electrolytes depend on their vapor pressure and thermal stability, have the potential to considerably influence the function of actuators. For instance, the near-zero vapor pressure and high thermal stability over a wide range of temperatures can noticeably enhance the actuator lifetime and also the operational temperature range.[234]

6.8.4 Adjustable-Speed Drive Ride-Through

Several electric devices, from home appliances to industrial plants, face a critical power disturbance problem caused by voltage sags and momentary interruptions. Voltage sags in general do not cause any damage in equipments, but can easily interrupt the operation of sensitive loads such as electronic adjustable speed drives (ASDs).[235] Voltage sags are the major cause for a temporary decrease voltage triggering an undervoltage trip leading to nuisance tripping of ASDs used in continuous-process industries, which contributes to a loss in overall income.[236] For instance, statistical data collected during the year 1977 showed that power quality disruptions costed U.S. companies more than $25 billion annually.[237,238] The nuisance of ASDs can be very disruptive to an industrial process while the voltage slumps 15%–20% under its nominal value.[239,240] Thus, a practical ride-through scheme for an ASD based on supercapacitor during voltage sag is necessarily required to maintain the ASD DC bus voltage under voltage sag condition.

A number of solutions have been raised to overcome this power tripping problem[241] including lead-acid batteries, flywheels, and ultracapacitors. Among these, batteries and flywheels seem to have profound maintenance, safety, and cost issues.[242,243] The supercapacitors, however, are the best candidates, as they require high power density and relatively low energy density. They provide a cost-effective solution at moderate power levels in regard to both initial and ownership costs.[241,244] One important thing that should be taken into account is that voltage decreases during supercapacitor discharge. To address this issue, Corley and coworkers designed and fabricated a 100-kW prototype ride-through system with a reliable, efficient, and cost-effective DC-DC convertor for Maxwell Technologies.[243] A DC-DC conversion system can adjust the bus voltage to allow for ASD ride-through while the supercapacitor voltage decreases during discharge. Their experimental results from using metal- or carbon-based supercapacitor proved the capability of the system to support ASDs and similar loads during the voltage sags and brief outages.

The patent from Enjeti and Duran-Gomez[245] introduced a new method for a ride-through system for ASD. They claimed that their invention, method, and system for ride-through of an ASD for voltage sags and short-term power interruptions substantially eliminates or reduces the disadvantages of any previous methods. They also stated that using an additional energy source diode like a supercapacitor, battery, fuel cell, photovoltaic cell, and flywheel can maintain the dc-link voltage value in the event of voltage sag, or even a complete loss of input power, energy source. The difference between their system with and the existing systems was the addition of three diodes, an inductor, and a control unit connected to dynamic braking circuit, which keeps the ride-through circuit activated when necessary.

In 2002, Duran-Gomez and coworkers simulated, designed, and fabricated a ride-through with flyback converter modules powered by supercapacitors.[246] Their simulation and experimental results from loaded and nonloaded ride-through systems demonstrated the feasibility of their proposed approach. The voltage of both systems was reduced from 450 V to 230 V by a programmable AC power source that facilitated the generation of a wide variety of transient and steady-state power quality disturbances. The nonloaded system showed the decay of 5 s for the DC-link voltage under short-term power interruptions, but the ride-through system effectively maintained the voltage at its nominal value.

6.9   Conclusion

In summary, the progress up to date on the supercapacitor electrode materials, including CNT-based, transition metal oxides and conductive polymers, has been investigated. CNTs, because of their extraordinary mechanical properties, high conductivity and surface area, good corrosion resistance, high temperature stability, and percolated pore structure, are the most promising carbonaceous material for supercapacitor application. Compared with conventional capacitors, a higher energy density and a corresponding cycle life can be achieved by hybrid electrochemical supercapacitors that benefit from both double-layer and pseudocapacitive properties. It has been shown that developing an energy storage device by combining the transition metal oxides or conductive polymers with CNTs, high electrical energy and power are achieved. Hence, fabrication of CNT-based nanocomposite materials and the use of such materials in supercapacitor electrodes provide a significant higher SC value than that from pure CNTs. RuO_2, having the best rectangular shape of CV curve and exhibiting the best capacitor behavior compared to other transition metal oxides, is the most promising pseudocapacitive electrode material and it gives a superior SC value. However, the toxicity and high cost of the precious metal (Ru) are major disadvantages for commercial production of RuO_2-based electrodes. MnO_2 and NiO are great alternatives for RuO_2 as they are the best electroactive materials among transition metal oxides with respect to both specific capacitance and cost effectiveness. Ceramics oxide (mixed metal oxide) from a combination of different transition metals using novel mixing methodologies will lead to superior pseudocapacitive behavior and high power-energy density. Ceramics are sometimes preferred to single-metal oxides, as adding one or more oxides helps in augmenting the capacitances because they are complementary in their properties. The advantages of using ceramics are in their high oxide ion conductivity and rapid charge transfer resulting in an increase of the SC value of supercapacitor. Conducting polymers due to their good electrical conductivity, large pseudocapacitance, and relatively low cost are the great candidate materials to be used with CNTs and ceramics in supercapacitor electrodes. Among the polymers, PEDOT seems to have a better cyclability than all other types of conducting polymers, and it is recommended to be used in hybrid supercapacitor materials. Hybrid supercapacitors are beneficial as they have high power density, long life expectancy, long shelf life, and a wide range of operating temperatures, and they are environmental friendly. On the other hand, a lot of research is under way to meet challenges like low energy density, high cost, and low operating voltage of the supercapacitor materials. Current applications of supercapacitors are in memory backup, electric vehicle power quality, battery improvement, electromechanical actuators, ADS ride-through, etc. Depending on the power density, supercapacitors are divided into two segments: (1) large cans from 100 F to 5000 F used for heavy-duty applications such as automotive industry and utilities and (2) small devices in the range of 0.1 F–10 F used in general electronics. Recently, quite large numbers of patents have reported the development and use of hybrid electrode materials in supercapacitors electrode materials and many companies fabricated hybrid supercapacitors in electronic devices. Active researches are still continuing on all aspects of hybrid supercapacitors, i.e., electrode materials (anodes, cathodes), electrolytes, and cell construction. The major factors limiting their wider application remain as their cost and safety and the recent developments from researchers have been taken place to address all of these limitations.

References

1. Mangematin V, Walsh S. The future of nanotechnologies. *Technovation* 2012; 32: 157–160.
2. Schilthuizen S, Simonis F. *Nanotechnology: Innovation Opportunities for Tomorrow's Defence.* TNO Science & Industry, 2006.
3. Filipponi L, Sutherland D. *Information and Communication Technologies (Ict), in Module₂-Application of Nanotechnology Interdiciplinary Nanoscience Center (iNano),* Denmark: Aarhus University, 2010.
4. Simonis F, Schilthuizen S. *Nanotechnology: Innovation Opportunities for Tomorrow's Defence.* TNO Science & Industry, 2006.
5. Thilagavathi G, Raja ASM, Kannaian T. Nanotechnology and protective clothing for defence personnel. *Def Sci J* 2008; 58: 451–459.
6. Aricò AS, Bruce P, Scrosati B, et al. Nanostructured materials for advanced energy conversion and storage devices. *Nat Mater* 2005; 4: 366–377.
7. Lu S, Corzine KA, Ferdowsi M. A new battery/ultracapacitor energy storage system design and its motor drive integration for hybrid electric vehicles. *IEEE Trans Veh Techno* 2007; 56: 1516–1523.
8. Kiamahalleh MV, Zein SHS, Najafpour G, et al. Multiwalled carbon nanotubes based nanocomposites for supercapacitors: A review of electrode materials. *Nano* 2012; 7: 29.
9. Halper MS, Ellenbogen JC. Supercapacitors: A brief overview. MITRE Nanosystems Group. Virginia: McLean, 2006.
10. Zhao X, Sánchez BM, Dobson PJ, et al. The role of nanomaterials in redox-based supercapacitors for next generation energy storage devices. *Nanoscale* 2011; 3: 839–855.
11. Becker HI. Low voltage electrolytic capacitor. United States: Gen Electric, 1957.
12. Boos DL. Electrolytic capacitor having carbon paste electrodes. United States: Standard Oil Co, 1970.
13. Conway BE. Transition from "supercapacitor" to "battery" behavior in electrochemical energy storage. *J Electrochem Soc* 1991; 138: 1539–1548.
14. Sarangapani S, Tilak BV, Chen CP. Materials for electrochemical capacitors. *J Electrochem Soc* 1996; 143: 3791–3799.
15. Murphy TC, Wright RB, Sutula RA. *Proceedings of the Symposium on Electrochemical Capacitors II* 1997; 96: 258.
16. Palacin MR, Guibert AD, Collins J, et al. *Strategic Energy Technology Plan.* Luxembourg: Publications Office of the European Union, 2011.
17. Conway BE. *Electrochemical Supercapacitor.* New York: Kluwer Academic/Plenum Publishers, 1999.
18. Namisnyk AM. A survey of electrochemical supercapacitor technology. Electrical Engineering. Sydney: University of Technology, 2003.
19. *Nanomaterials State of the Market Q₃: Stealth Success, Broad Impact.* LUX Research, 2008.
20. Inventor of Enable IPC's ultracapacitor technology to present at ISEE/CAP 09. AZoNano, 2009.
21. Liu T, Vijayendran BR, Gupta A, et al. *Carbon Nanotube Nanocomposites, Methods of Making Carbon Nanotube Nanocomposites, and Devices Comprising the Nanocomposites.* United States: Battelle Memorial Institute, 2011.

22. Nuintek. Comparison of capacitor, supercapacitor and battery. 2006. Available from: http://www.nuin.co.kr.
23. Kötz R, Carlen M. Principles and applications of electrochemical capacitors. *Electrochim Acta* 2000; 45: 2483–2498.
24. Schneuwly A. Designing powerful electronic solutions with ultracapacitors. Switzerland: Maxwell Technologies SA: Rossens, 2006.
25. Celzard A, Collas F, Maréché JF, et al. Porous electrodes-based double-layer supercapacitors: Pore structure versus series resistance. *J Power Sources* 2002; 108: 153–162.
26. Gamby J, Taberna PL, Simon P, et al. Studies and characterisations of various activated carbons used for carbon/carbon supercapacitors. *J Power Sources* 2001; 101: 109–116.
27. Xie Y, Zhou L, Huang C, et al. Fabrication of nickel oxide-embedded titania nanotube array for redox capacitance application. *Electrochim Acta* 2008; 53: 3643–3649.
28. Feng G, Cummings PT. Supercapacitor capacitance exhibits oscillatory behavior as a function of nanopore size. *J Phys Chem Lett* 2011; 2: 2859–2864.
29. Xie K, Qin X, Wang X, et al. Carbon nanocages as supercapacitor electrode materials. *Adv Mater* 2012; 24: 347–352.
30. Yamada Y, Tanaike O, Liang TT, et al. Electric double layer capacitance performance of porous carbons prepared by defluorination of polytetrafluoroethylene with potassium. *Electrochem Solid State Lett* 2002; 5: A283–A285.
31. An KH, Kim WS, Park YS, et al. Electrochemical properties of high-power supercapacitors using single-walled carbon nanotube electrodes. *Adv Funct Mater* 2001; 11: 387–392.
32. Frackowiak E, Metenier K, Bertagna V, et al. Supercapacitor electrodes from multiwalled carbon nanotubes. *Appl Phys Lett* 2000; 77: 2421–2423.
33. Ko JM, Kim KM. Electrochemical properties of MnO_2/activated carbon nanotube composite as an electrode material for supercapacitor. *Mater Chem Phys* 2009; 114: 837–841.
34. Amitha FE, Leela A, Reddy M, et al. A non-aqueous electrolyte-based asymmetric supercapacitor with polymer and metal oxide/multiwalled carbon nanotube electrodes. *J Nanopart Res* 2009; 11: 725–729.
35. Lee HY, Goodenough JB. Ideal supercapacitor behavior of amorphous $V_2O_5 \cdot NH_2O$ in potassium chloride (KCl) aqueous solution. *J Solid State Chem* 1999; 148: 81–84.
36. Jiang J, Kucernak A. electrochemical supercapacitor material based on manganese oxide: Preparation and characterization. *Electrochim Acta* 2002; 47: 2381–2386.
37. Stotz S, Wagner C. Die Loslichkeit von Wasserdampf und Wasserstoff in Festen Oxiden. Ber. Bunsenges. *Phys Chem* 1966; 70: 781–788.
38. Takahashi T, Iwahara H. Solid-state ionics: Protonic conduction in perovskite type oxide solid solutions. *Revue de Chimie Minerale* 1980; 17: 243–253.
39. Browall KW, Muller O, Doremus RH. Oxygen ion conductivity in oxygen deficient perovskite-related oxides. *Mater Res Bull* 1976; 11: 1475–1482.
40. Iwahara H, Uchida H, Tanaka S. High-temperature type proton conductor based on $SrCeO_3$ and its application to solid electrolyte fuel cells. *Solid State Ionics* 1983; 9–10: 1021–1025.

41. Iwahara H, Uchida H, Ono K, et al. Proton conduction in sintered oxides based on BaCeO3. *J Electrochem Soc* 1988; 135: 529–533.
42. Arbizzani C, Mastragostino M, Soavi F. New trends in electrochemical supercapacitors. *J Power Sources* 2001; 100: 164–170.
43. Chen WC, Wen TC, Teng H. Polyaniline-deposited porous carbon electrode for supercapacitor. *Electrochim Acta* 2003; 48: 641–649.
44. Frackowiak E, Jurewicz K, Szostak K, et al. Nanotubular materials as electrodes for supercapacitors. *Fuel Process Technol* 2002; 77–78: 213–219.
45. Hendriks MGHM, Heijman MJGW, Van Zyl WE, et al. Solid state supercapacitor materials: Layered structures of yttria-stabilized zirconia sandwiched between platinum/yttria-stabilized zirconia composites. *J Appl Phys* 2001; 90: 5303–5307.
46. Lee H, Kim H, Cho MS, et al. Fabrication of polypyrrole (PPY)/carbon nanotube (CNT) composite electrode on ceramic fabric for supercapacitor applications. *Electrochim Acta* 2011; 56: 7460–7466.
47. Yan S, Wang H, Qu P, et al. RuO_2/Carbon nanotubes composites synthesized by microwave-assisted method for electrochemical supercapacitor. *Synth.Met* 2009; 159: 158–161.
48. Mrunal AK, Youngjin C, Pritesh H, et al. Electrical actuation and readout in a nanoelectromechanical resonator based on a laterally suspended zinc oxide nanowire. *Nanotechn* 2012; 23: 025501.
49. Gibson C, Karthikeyan A. Electrical energy storage device containing an electroactive separator. United States: Wisys Technology Foundation, Inc., 2011.
50. Burke A. Ultracapacitors: Why, how, and where is the technology? *J Power Sources* 2000; 91: 37–50.
51. Rafik F, Gualous H, Gallay R, et al. Frequency, thermal and voltage supercapacitor characterization and modeling. *J Power Sources* 2007; 165: 928–934.
52. Wu HP, He DW, Wang YS, et al. Graphene as the electrode material in supercapacitors. in *8th International Vacuum Electron Sources Conference and Nanocarbon (IVESC)*, 2010.
53. Stoller MD, Ruoff RS. Best practice methods for determining an electrode material's performance for ultracapacitors. *Energ Environ Sci* 2010; 3: 1294–1301.
54. Zhang LL, Zhao XS. Carbon-based materials as supercapacitor electrodes. *Chem. Soc Rev* 2009; 38: 2520–2531.
55. Pan H, Li J, Feng Y. Carbon nanotubes for supercapacitor. *Nanoscale Res Lett* 2010; 5: 654–668.
56. Inagaki M, Konno H, Tanaike O. Carbon materials for electrochemical capacitors. *J Power Sources* 2010; 195: 7880–7903.
57. Wang G, Zhang L, Zhang J. A review of electrode materials for electrochemical supercapacitors. *Chem Soc Rev* 2012; 41: 797–828.
58. Kisacikoglu MC, Uzunoglu M, Alam MS. Load sharing using fuzzy logic control in a fuel cell/ultracapacitor hybrid vehicle. *Int J Hydrogen Energy* 2009; 34: 1497–1507.
59. Ma SB, Nam KW, Yoon WS, et al. A novel concept of hybrid capacitor based on manganese oxide materials. *Electrochem Commun* 2007; 9: 2807–2811.
60. Edwards CJC, Hitchen DA, Sharples M. Porous carbon structures and methods for their preparation. Internationale Octrooi Maatschappij, U.S. Patent 4775655 A, 1988.

61. Mastragostino M, Arbizzani C, Soavi F. Polymer-based supercapacitors. *J Power Sources* 2001; 97–98: 812–815.
62. Gujar TP, Kim WY, Puspitasari I, et al. Electrochemically deposited nanograin ruthenium oxide as a pseudocapacitive electrode. *Int J Electrochem Sci* 2007; 2: 666–673.
63. Liu T, Pell WG, Conway BE. Self-discharge and potential recovery phenomena at thermally and electrochemically prepared RuO_2 supercapacitor electrodes. *Electrochim Acta* 1997; 42: 3541–3552.
64. Michell D, Rand DAJ, Woods R. Analysis of the anodic oxygen layer on iridium by x-ray emission, electron diffraction and electron microscopy. *J Electroanal Chem* 1977; 84: 117–126.
65. Liu DQ, Yu SH, Son SW, et al. Electrochemical performance of iridium oxide thin film for supercapacitor prepared by radio frequency magnetron sputtering method. ECS Transactions, 2008; 16: 103–109.
66. Micka K, Zábranský Z. Study of iron oxide electrodes in an alkaline electrolyte. *J Power Sources* 1987; 19: 315–323.
67. Nagarajan N, Zhitomirsky I. Cathodic electrosynthesis of iron oxide films for electrochemical supercapacitors. *J Appl Electrochem* 2006; 36: 1399–1405.
68. Du X, Wang C, Chen M, et al. Electrochemical performances of nanoparticle Fe_3O_4/activated carbon supercapacitor using KOH electrolyte solution. *J Phys Chem C* 2009; 113: 2643–2646.
69. Deng MG. Investigation of anion reducing agents prepared mesoporous MnO_2 for supercapacitors. *Gongneng Cailiao/J Func Mater* 2008; 39: 2002–2004.
70. Goodenough JB. Manganese oxides as battery cathodes. *Electrochemical Society Extended Abstracts*, 1984.
71. Ragupathy P, Park DH, Campet G, et al. Remarkable capacity retention of nano-structured manganese oxide upon cycling as an electrode material for supercapacitor. *J Phys Chem C* 2009; 113: 6303–6309.
72. Patil UM, Salunkhe RR, Gurav KV, et al. Chemically deposited nanocrystalline nio thin films for supercapacitor application. *Appl Surf Sci* 2008; 225: 2603–2607.
73. Wang YG, Xia YY. Electrochemical capacitance characterization of NiO with ordered mesoporous structure synthesized by template Sba-15. *Electrochim Acta* 2006; 51: 3223–3227.
74. Zheng YZ, Ding HY, Zhang MI. Preparation and electrochemical properties of nickel oxide as a supercapacitor electrode material. *Mater Res Bull* 2009; 44: 403–407.
75. Laforgue A, Simon P, Sarrazin C, et al. Polythiophene-based supercapacitors. *J Power Sources* 1999; 80: 142–148.
76. Sharma RK, Rastogi AC, Desu SB. Pulse polymerized polypyrrole electrodes for high energy density electrochemical supercapacitor. *Electrochem Commun* 2008; 10: 268–272.
77. Bélanger D, Ren X, Davey J, et al. Characterization and long-term performance of polyaniline-based electrochemical capacitors. *J Electrochem Soc* 2000; 147: 2923–2929.
78. Yahya AI, Jonho C, Ryon SS, et al. Hydrogel-assisted polyaniline microfiber as controllable electrochemical actuatable supercapacitor. *J Electrochem Soc* 2009; 156: A313–A317.
79. Fang B, Binder L. Enhanced surface hydrophobisation for improved performance of carbon aerogel electrochemical capacitor. *Electrochim Acta* 2007; 52: 6916–6921.

80. Frackowiak E, Béguin F. Carbon materials for the electrochemical storage of energy in capacitors. *Carbon* 2001; 39: 937–950.
81. Obreja VVN. On the performance of supercapacitors with electrodes based on carbon nanotubes and carbon activated material-A review. *Phys E: Low-dimens Sys Nanostruct* 2008; 40: 2596–2605.
82. Shi H. Activated carbons and double layer capacitance. *Electrochim Acta* 1996; 41: 1633–1639.
83. Diederich L, Barborini E, Piseri P, et al. Supercapacitors based on nanostructured carbon electrodes grown by cluster-beam deposition. *Appl Phys Lett* 1999; 75: 2662–2664.
84. He X, Jiang L, Yan S, et al. Direct synthesis of porous carbon nanotubes and its performance as conducting material of supercapacitor electrode. *Diamond Relat Mater* 2008; 17: 993–998.
85. Maletin Y, Strizhakova N, Izotov Z, et al. Supercapacitors: Old problems and new trends. *New Carbon Based Materials for Electrochemical Energy Storage Systems: Batteries, Supercapacitors and Fuel Cells.* Argonne, Illinois, U.S.A., pp. 51–62, 2006.
86. Niu C, Sichel EK, Hoch R, et al. High power electrochemical capacitors based on carbon nanotube electrodes. *Appl Phys Lett* 1997; 70: 1480–1482.
87. Qu D, Shi H. Studies of activated carbons used in double-layer capacitors. *J Power Sources* 1998; 74: 99–107.
88. Randin JP, Yeager E. Differential capacitance study on the basal plane of stress-annealed pyrolytic graphite. *J Electroanal Chem* 1972; 36: 257–276.
89. Randin JP, Yeager E. Differential capacitance study on the edge orientation of pyrolytic graphite and glassy carbon electrodes. *J Electroanal Chem* 1975; 58: 313–322.
90. Rodríguez-Reinoso F, Molina-Sabio M, Activated carbons from lignocellulosic materials by chemical and/or physical activation: An overview. *Carbon* 1992; 30: 1111–1118.
91. Torregrosa R, Martín-Martínez J. Activation of lignocellulosic materials: A comparison between chemical, physical and combined activation in terms of porous texture. *Fuel* 1991; 70: 1173–1180.
92. Hatori H, Yamada Y, Shiraishi M. Preparation of macroporous carbon films from polyimide by phase inversion method. *Carbon* 1992; 30: 303–304.
93. Barby D, Haq Z. European Patent 0060138 1982. Assigned to Unilever
94. Pekala RW, Alviso CT, LeMay JD. Organic aerogels: Microstructural dependence of mechanical properties in compression. *J Non-Cryst Solids* 1990; 125: 67–75.
95. Ryoo R, Joo SH. Nanostructured carbon materials synthesized from mesoporous silica crystals by replication, in *Stud. Surf. Sci Catal* 2004; 148: 241–260.
96. Jurewicz K, Vix-Guterl C, Frackowiak E. et al. Capacitance properties of ordered porous carbon materials prepared by a templating procedure. *J Phys Chem Solids* 2004; 65: 287–293.
97. Lee J, Han S, Hyeon T. Synthesis of new nanoporous carbon materials using nanostructured silica materials as templates. *J Mater Chem* 2004; 14: 478–486.
98. Vix-Guterl C, Saadallah S, Jurewicz K, et al. Supercapacitor electrodes from new ordered porous carbon materials obtained by a templating procedure. *Mater Sci Eng B* 2004; 108: 148–155.
99. Bonnefoi L, Simon P, Fauvarque JF, et al. Electrode compositions for carbon power supercapacitors. *J Power Sources* 1999; 80: 149–155.

100. Rose MF, Johnson C, Owens T, et al. Limiting factors for carbon-based chemical double-layer capacitors. *J Power Sources* 1994; 47: 303–312.
101. Mi H, Zhang X, An S, et al. Microwave-assisted synthesis and electrochemical capacitance of polyaniline/multi-wall carbon nanotubes composite. *Electrochem Commun* 2007; 9: 2859–2862.
102. Osaka T, Liu X, Nojima M, et al. Electrochemical double layer capacitor using an activated carbon electrode with gel electrolyte binder. *J Electrochem Soc* 1999; 146: 1724–1729.
103. Reddy ALM, Ramaprabhu S. Nanocrystalline metal oxides dispersed multi-walled carbon nanotubes as supercapacitor electrodes. *J Phys Chem C* 2007; 111: 7727–7734.
104. Beck F, Dolata M. Fluorine-free binders for carbon black based electrochemical supercapacitors. *J Appl Electrochem* 2001; 31: 517–521.
105. Lin C, Ritter JA. Effect of synthesis pH on the structure of carbon xerogels. *Carbon* 1997; 35: 1271–1278.
106. Zanto EJ, Al-Muhtaseb SA, Ritter JA. Sol-gel-derived carbon aerogels and xerogels: Design of experiments approach to materials synthesis. *Ind Eng Chem Res* 2002; 41: 3151–3162.
107. Chu X, Kinoshita K. Electrochemical capacitors. *Electrochem Soc Proc Ser* 1996; 235–245.
108. Peng C, Zhang S, Jewell D, et al. Carbon nanotube and conducting polymer composites for supercapacitors. *Prog Nat Sci* 2008; 18: 777–788.
109. Arbizzani C, Mastragostino M, Meneghello L, et al. Electronically conducting polymers and activated carbon: Electrode materials in supercapacitor technology. *Adv Mater* 1996; 8: 331–334.
110. Balducci A, Wesley AH, Marina M, et al. Cycling stability of a hybrid activated carbon/poly(3-Methylthiophene) supercapacitor with N-butyl-N-methylpyrrolidinium bis(trifluoromethanesulfonyl)imide ionic liquid as electrolyte. *Electrochim Acta* 2005; 50: 2233–2237.
111. Fan LZ, Maier J. High-performance polypyrrole electrode materials for redox supercapacitors. *Electrochem l Commun* 2006; 8: 937–940.
112. Fonseca CP, Benedetti JOE, Neves S. Poly(3-methyl thiophene)/pvdf composite as an electrode for supercapacitors. *J Power Sources* 2006; 158: 789–794.
113. Ren X, Gottesfeld S, Ferraris JP. In *Proceedings of the Symposium on Supercapacitors* 95-29. 1996. Pennington, NJ. p. 138.
114. Zhou C, Kumar S, Doyle CD, et al. Functionalized single wall carbon nanotubes treated with pyrrole for electrochemical supercapacitor membranes. *Chem Mater* 2005; 17: 1997–2002.
115. Wu M, Shuren Z. Electrochemical capacitance of mwcnt/polyaniline composite coatings grown in acidic MWCNT suspensions by microwave-assisted hydrothermal digestion. *Nanotechnol* 2007; 18: 385603.
116. Wang J, Xu Y, Chen X, et al. Capacitance properties of single wall carbon nanotube/polypyrrole composite films. *Compos Sci Technol* 2007; 67: 2981–2985.
117. Chen L, Yuan C, Dou H, et al. Synthesis and electrochemical capacitance of core-shell poly (3,4-ethylenedioxythiophene)/poly (sodium 4-styrenesulfonate)-modified multiwalled carbon nanotube nanocomposites. *Electrochim Acta* 2009; 54: 2335–2341.
118. Lao ZJ. Metal oxides as electrode materials for electrochemical capacitors. Institute for superconducting and electronic materials. University of Wollongong, 2006.

119. Su YZ, Niu YP, Xiao YZ, et al. Novel conducting polymer poly[bis(phenylamino) disulfide]: Synthesis, characterization, and properties. *J Polym Sci Part A: Polym Chem* 2004; 42: 2329–2339.

120. Prasad KR, Munichandraiah N. Fabrication and evaluation of 450 F electrochemical redox supercapacitors using inexpensive and high-performance, polyaniline coated, stainless-steel electrodes. *J Power Sources* 2002; 112: 443–451.

121. Gupta V, Miura N. High performance electrochemical supercapacitor from electrochemically synthesized nanostructured polyaniline. *Mater Lett* 2006; 60: 1466–1469.

122. Xu Y, Wang J, Sun W. Capacitance properties of poly(3,4-ethylenedioxythiophene)/polypyrrole composites. *J Power Sources* 2006; 159: 370–373.

123. Zhao DD, Bao SJ, Zhou WJ, et al. Preparation of hexagonal nanoporous nickel hydroxide film and its application for electrochemical capacitor. *Electrochem Commun* 2007; 9: 869–874.

124. Trasatti S, Buzzanca G. Ruthenium dioxide: A new interesting electrode material solid states structure and electrochemical behaviour. *J Electroanal Chem* 1971; 29: A1–A5.

125. Zheng JP, Cygan PJ, Jow TR. Hydrous ruthenium oxide as an electrode material for electrochemical capacitors. *J Electrochem Soc* 1995; 142: 2699–2703.

126. Ardizzone S, Fregonara G, Trasatti S. "Inner" and "outer" active surface of RuO$_2$ electrodes. *Electrochim Acta* 1990; 35: 263–267.

127. Hang BT, Hayashi H, Yoon SH, et al. Fe$_2$O$_3$-filled carbon nanotubes as a negative electrode for an Fe-Air battery. *J Power Sources* 2008; 178: 393–401.

128. Wu NL. Nanocrystalline oxide supercapacitors. *Mater Chem Phys* 2002; 75: 6–11.

129. Wu GT, Wang CS, Zhang XB, et al. Lithium insertion into CuO/carbon nanotubes. *J Power Sources* 1998; 75: 175–179.

130. Wu HQ, Wei XW, Shao MW, et al. Synthesis of copper oxide nanoparticles using carbon nanotubes as templates. *Chem Phys Lett* 2002; 364: 152–156.

131. Toupin M, Brousse T, Bélanger D. Influence of microstucture on the charge storage properties of chemically synthesized manganese dioxide. *Chem Mater* 2002; 14: 3946–3952.

132. Xia H, Xiao W, Lai MO, et al. Facile synthesis of novel nanostructured MnO$_2$ thin films and their application in supercapacitors. *Nanoscale Res Lett* 2009; 1–6.

133. Wang XL, Zhang XG, He KX. Hydrothermal synthesis of NiO microspheres and their electrochemical properties. *Gongneng cailiao/J Func Mater* 2006; 37: 355–357.

134. Zheng YZ, Zhang ML. Preparation and electrochemical properties of nickel oxide by molten-salt synthesis. *Mater Lett* 2007; 61: 3967–3969.

135. Guther TJ, Oesten R, Garche J. Development of supercapacitor materials based on perovskites. *Proceedings of the Symposium on Electrochemical Capacitors (II)*. 1997. Pennington, NJ. p. 16–25.

136. Liu TC, Pell WG, Conway BE, et al. Behavior of molybdenum nitrides as materials for electrochemical capacitors comparison with ruthenium oxide. *J.Electrochem Soc* 1998; 145: 1882–1888.

137. Malak A, Fic K, Lota G, et al. Hybrid materials for supercapacitor application. *J Solid State Electrochem* 2010; 14: 811–816.

138. Yue H, Huang X, Yang Y. Preparation and electrochemical performance of manganese oxide/carbon nanotubes composite as a cathode for rechargeable lithium battery with high power density. *Mater Lett* 2008; 62: 3388–3390.

139. Lee JY, Liang K, An KH, et al. Nickel oxide/carbon nanotubes nanocomposite for electrochemical capacitance. *Synth Met* 2005; 150: 153–157.
140. Nam KW, Kim KH, Lee ES, et al. Pseudocapacitive properties of electrochemically prepared nickel oxides on 3-dimensional carbon nanotube film substrates. *J Power Sources* 2008; 182: 642–652.
141. Wei TY, Chen CH, Chang KH, et al. Cobalt oxide aerogels of ideal supercapacitive properties prepared with an epoxide synthetic route. *Chem Mater* 2009; 21: 3228–3233.
142. Xiong S, Yuan C, Zhang X, et al. Controllable synthesis of mesoporous Co_3O_4 nanostructures with tunable morphology for application in supercapacitors. *Chem-A Eur J* 2009; 15: 5320–5326.
143. Patake VD, Joshi SS, Lokhande CD, et al. Electrodeposited porous and amorphous copper oxide film for application in supercapacitor. *Mater Chem Phys.* 2009; 114: 6–9.
144. Cottineau T, Toupin M, Delahaye T, et al. Nanostructured transition metal oxides for aqueous hybrid electrochemical supercapacitors. *Appl Phys A: Mater Sci Process* 2006; 82: 599–606.
145. Wu NL, Lan YP, Han CY, et al. Electrochemical capacitor and hybrid power sources. *Electrochem Soc Proc Ser* 2002; 95–106.
146. Cheng L, Li HQ, Xia YY. A hybrid nonaqueous electrochemical supercapacitor using nano-sized iron oxyhydroxide and activated carbon. *J Solid State Electrochem* 2006; 10: 405–410.
147. Park GJ, Kalpana D, Thapa AK, et al. A novel hybrid supercapacitor using a graphite cathode and a niobium(V) oxide anode. *Bull Korean Chem Soc* 2009; 30: 817–820.
148. Hughes M, Chen GZ, Shaffer MSP, et al. Electrochemical capacitance of a nanoporous composite of carbon nanotubes and polypyrrole. *Chem Mater* 2002; 14: 1610–1613.
149. Konyushenko EN, Stejskal J, Trchová M, et al. Multi-wall carbon nanotubes coated with polyaniline. *Polymer* 2006; 47: 5715–5723.
150. Kumar SA, Chen SM. Electroanalysis of NADH using conducting and redox active polymer/carbon nanotubes modified electrodes—A review. *Sens* 2008; 8: 739–766.
151. Lee YH, An KH, Lim SC, et al. Applications of carbon nanotubes to energy storage devices. *New Diamond Front Carbon Technol* 2002; 12 : 209–228.
152. Xiao Q, Zhou X. The study of multiwalled carbon nanotube deposited with conducting polymer for supercapacitor. *Electrochim Acta* 2003; 48: 575–580.
153. Laforgue A, Simon P, Fauvarque JF, et al. Activated carbon/conducting polymer hybrid supercapacitors. *J Electrochem Soc* 2003; 150: A645–A651.
154. Shekhar S, Prasad V, Subramanyam SV. Transport properties of conducting amorphous carbon-poly(vinyl chloride) composite. *Carbon* 2006; 44: 334–340.
155. Snook GA, Wilson GJ, Pandolfo AG. Mathematical functions for optimisation of conducting polymer/activated carbon asymmetric supercapacitors. *J Power Sources* 2009; 186: 216–223.
156. Emmenegger C, Mauron P, Sudan P, et al. Investigation of electrochemical double-layer (ECDL) capacitors electrodes based on carbon nanotubes and activated carbon materials. *J Power Sources* 2003; 124: 321–329.
157. Huang CW, Chuang CM, Ting JM. Significantly enhanced charge conduction in electric double layer capacitors using carbon nanotube-grafted activated carbon electrodes. *J Power Sources* 2008; 183: 406–410.

158. Huang CW, Teng H. Influence of carbon nanotube grafting on the impedance behavior of activated carbon capacitors. *J Electrochem Soc* 2008; 155.
159. Taberna PL, Chevallier G, Simon P, et al. Activated carbon-carbon nanotube composite porous film for supercapacitor applications. *Mater Res Bull* 2006; 41: 478–484.
160. Wang XF, Liang J, Wang DZ. Electrochemical performance of combined Ni(OH)2/carbon nanotubes (CNTs) composite-activated carbon supercapacitor. *Chin J Inorg Chem* 2005; 21: 35–42.
161. Jang J, Han S, Hyeon T, et al. Electrochemical capacitor performance of hydrous ruthenium oxide/mesoporous carbon composite electrodes. *J Power Sources* 2003; 123; 79–85.
162. Liang YY, Li HL, Zhang XG, A novel asymmetric capacitor based on Co(OH)2/USY composite and activated carbon electrodes. *Mater Sci Eng A* 2008; 473: 317–322.
163. Wu M, Snook GA, Chen GZ, et al. Redox deposition of manganese oxide on graphite for supercapacitors. *Electrochem Commun* 2004; 6: 499–504.
164. Zhang L, Song JY, Zou JY, et al. $RuO_2 \cdot Xh_2O/Ac$ composite electrode and properties of super-capacitors. *Wuji Cailiao Xuebao/J Inorg Mater* 2005; 20: 745–749.
165. Kiamahalleh MV, Sata SA, Buniran S, et al. Remarkable stability of supercapacitor material synthesized by manganese oxide filled in multiwalled carbon nanotubes. *Curr Nanosci* 2010; 6: 553–559.
166. Kiamahalleh MV, Sata SA, Surani B, et al. A comparative study on the electrochemical performance of nickel oxides and manganese oxides nanocomposites based multiwall carbon nanotubes. *World Appl Sci J* 2009; 6: 711–718.
167. Yan J, Zhou H, Yu P, et al. A general electrochemical approach to deposition of metal hydroxide/oxide nanostructures onto carbon nanotubes. *Electrochem Commun* 2008; 10: 761–765.
168. Fan Z, Chen J, Cui K, et al. Preparation and capacitive properties of cobalt-nickel oxides/carbon nanotube composites. *Electrochim Acta* 2007; 52: 2959–2965.
169. Li Q, Li KX, Gu JY, et al. Preparation and electrochemical characterization of cobalt-manganese oxide as electrode materials for electrochemical capacitors. *J Phys Chem Solids* 2008; 69; 1733–1739.
170. NuLi Y, Zhang P, Liu H, et al. Nickel-cobalt oxides/carbon nanoflakes as anode materials for lithium-ion batteries. *Mater Res Bull* 2009; 44: 140–145.
171. Hou Y, Cheng Y, Hobson T, et al. Design and synthesis of hierarchical MnO_2 nanospheres/carbon nanotubes/conducting polymer ternary composite for high performance electrochemical electrodes. *Nano Lett* 2010; 10: 2727–2733.
172. Kiamahalleh MV, Cheng CI, Sata SA, et al. Preparation and capacitive properties of nickel-manganese oxides/multiwalled carbon nanotube/poly (3,4-ethylene-dioxythiophene) composite material for electrochemical supercapacitor. *J Appl Sci* 2011; 11: 2346–2351.
173. Kiamahalleh MV, Cheng CI, Sata SA, et al. Highly efficient hybrid supercapacitor material from nickel-manganese oxides/MWCNT/PEDOT nanocomposite. *Nano* 2010; 5: 143–148.
174. Sivakkumar S, Ko JM, Kim DY, et al. Performance evaluation of CNT/polypyrrole/MnO_2 composite electrodes for electrochemical capacitors. *Electrochim Acta* 2007; 52: 7377–7385.

175. Sot MABM, Kiamahalleh MV, Najafpour G, et al. Optimization of specific capacitance for hybrid supercapacitor material based on nickel-manganese oxides/multiwalled carbon nanotubes/poly (3,4-ethylenedioxythiophene) using response surface methodology. *World Appl Sci J* (Special Issue of Carbon Nanotubes)2010; 9: 14–20.

176. Zamri MFMA, Zein SHS, Abdullah AZ, et al. The optimization of electrical conductivity using central composite design for polyvinyl alcohol/multiwalled carbon nanotube-manganese dioxide nanofiber composites synthesised by electrospinning. *J Appl Sci* 2012; 12: 345–353.

177. Lota K, Sierczynska A, Lota G. Supercapacitors based on nickel oxide/carbon materials composites. *Int J Electrochem* 2011; 2011: 1–6.

178. Kusko A, DeDad J. Stored energy-short-term and long-term energy storage methods. *IEEE Ind Appl Mag* 2007; 13: 66–72.

179. Uzunoglu M, Alam MS. Modeling and analysis of an Fc/Uc hybrid vehicular power system using a novel-wavelet-based load sharing algorithm. *IEEE Trans Energy Conver* 2008; 23: 263–272.

180. Pasquier AD, Plitz I, Gural J, et al. Characteristics and performance of 500 F asymmetric hybrid advanced supercapacitor prototypes. *J Power Sources* 2003; 113: 62–71.

181. Laforgue A, Simon P, Fauvarque JF, et al. Hybrid supercapacitors based on activated carbons and conducting polymers. *J Electrochem Soc* 2001; 148: A1130–A1134.

182. Du Pasquier A, Laforgue A, Simon P, et al. A nonaqueous asymmetric hybrid $Li_4Ti_5O_{12}$/poly(fluorophenylthiophene) energy storage device *J Electrochem Soc* 2002; 149: A302–A306.

183. Zhang Y, Feng H, Wu X, et al. Progress of electrochemical capacitor electrode materials: A review. *Int J Hydrogen Energy* 2009; 34: 4889–4899.

184. Chang JK, Lee MT, Cheng CW, et al. Pseudocapacitive behavior of Mn oxide in aprotic 1-ethyl-3-methylimidazolium-dicyanamide ionic liquid. *J Mater Chem* 2009; 19: 3732–3738.

185. Chen CY, Chien TC, Chan YC, et al. Pseudocapacitive properties of carbon nanotube/manganese oxide electrode deposited by electrophoretic deposition. *Diamond Relat Mater* 2009; 18: 482–485.

186. Largeot C, Portet C, Chmiola J, et al. Relation between the ion size and pore size for an electric double-layer capacitor. *J Am Chem Soc* 2008; 130: 2730–2731.

187. Balducci A, Bardi U, Caporali S, et al. Ionic liquids for hybrid supercapacitors. *Electrochem Commun* 2004; 6: 566–570.

188. Ruifeng Z, Chuizhou M, Feng Z, et al. High-performance supercapacitors using a nanoporous current collector made from super-aligned carbon nanotubes. *Nanotechnol* 2010; 21: 345701.

189. Yan J, Wei T, Shao B, et al. Electrochemical properties of graphene nanosheet/carbon black composites as electrodes for supercapacitors. *Carbon* 2010; 48: 1731–1737.

190. Sheng K, Sun Y, Li C, et al. Ultrahigh-rate supercapacitors based on eletrochemically reduced graphene oxide for AC line-filtering. *Sci Rep* 2012; 2: 1–5.

191. Bose S, Kuila T, Mishra AK, et al. Carbon-based nanostructured materials and their composites as supercapacitor electrodes. *J Mater Chem* 2012; 22: 767–784.

192. Li L, Qin ZY, Wang LF, et al. Anchoring alpha-manganese oxide nanocrystallites on multi-walled carbon nanotubes as electrode materials for supercapacitor. *J Nanopart Res* 2010; 12: 2349–2353.

193. Lu P, Xue D, Yang H, et al. Supercapacitor and nanoscale research towards electrochemical energy storage. *Int J Smart Nano Mater* 2012; 1–25.
194. Lu W, Hartman R, Qu L, et al. Nanocomposite electrodes for high-performance supercapacitors. *J Phys Chem Lett* 2011; 2: 655–660.
195. Yuan C, Hou L, Li D, et al. Enhanced supercapacitance of hydrous ruthenium oxide/mesocarbon microbeads composites toward electrochemical capacitors. *Int J Electrochem* 2012; 2012: 1–7.
196. Zheng JP, Jow TR. A new charge storage mechanism for electrochemical capacitors. *J Electrochem Soc* 1995; 142: L6–L8.
197. Ervin MH. *Carbon Nanotube and Graphene Based Supercapacitors: Rationale, Status, and Prospects*. Adelphi, 2010.
198. Shukla AK, Banerjee A, Ravikumar MK, et al. Electrochemical capacitors: Technical challenges and prognosis for future markets. *Electrochim Acta* 2012; 84: 165–173.
199. Matsumoto T, Komatsu T, Arai K, et al. Reduction of Pt usage in fuel cell electrocatalysts with carbon nanotube electrodes. *Chem Commun* 2004; 7: 840–841.
200. Jiang H, Li C, Sun T, et al. High-performance supercapacitor material based on $Ni(OH)_2$ nanowire-MnO_2 nanoflakes core-shell nanostructures. *Chem Commun* 2012; 48: 2606–2608.
201. Hu ZA, Xie YL, Wang YX, et al. Synthesis and electrochemical characterization of mesoporous CoxNi1−X layered double hydroxides as electrode materials for supercapacitors. *Electrochim Acta* 2009; 54: 2737–2741.
202. Zhang J, Kong LB, Cai JJ, et al. Hierarchically porous nickel hydroxide/mesoporous carbon composite materials for electrochemical capacitors. *Micropor Mesopor Mater* 2010; 132: 154–162.
203. Endo M, Takeda T, Kim YJ, et al. High power electric double layer capacitor (EDLC's); from operating principle to pore size control in advanced activated carbons. *Carbon sci* 2001; 1: 117–128.
204. Mars P. *A Survey of Supercapacitors, Their Applications, Power Design with Supercapacitors, and Future Directions*. Hong Kong, 2011.
205. Rightmire RA. *Electrical Energy Storage Apparatus*. U. S.: Patent, Editor, 1966.
206. P. Mars. Cap-Xx supercapacitors provide power backup for solid state drives. *White Pap* 2008; 1–18.
207. Drew J. Supercapacitor-based power backup system protects volatile data in handhelds when power is lost, in design note 498. 2011. Available from: http://www.linear.com/.
208. Lohner A, Evers W. Intelligent power management of a supercapacitor based hybrid power train for light-rail vehicles and city busses, in power electronics specialists conference. *PESC 04. 2004 IEEE 35th Annual*. 2004: 672–676.
209. Zorpette G. Super charged. *IEEE Spectrum*: 32–37, 2005.
210. Smith SC, Sen PK. Ultracapacitors and energy storage: Applications in electrical power system, in power symposium. *NAPS '08. 40th North American. 2008*: 1–6.
211. Maxwell Technologies. 2008. Available from: http://www.maxwell.com.
212. Beale SR, Gerson R, the use of ultracapacitors as the sole power plant in an autonomous, electric rail-guided vehicle, in applied power electronics conference and exposition. *APEC '04. Nineteenth Annual IEEE*. 2004: 1178–1183.
213. Rufer A, Hotellier D, Barrade P. A supercapacitor-based energy storage substation for voltage compensation in weak transportation networks. *Power Deliv IEEE Trans* 2004; 19: 629–636.

214. Barker PP. Ultracapacitors for use in power quality and distributed resource applications. *Power Engineering Society Summer Meeting*, IEEE. 2002: 316–320.
215. Ackerman MC, Jefferson CM. *Global Emissions Due to Urban Transport and the Potential for Their Reduction*, Borrego C, Sucharov L, Editors. Computational Mechanics Publ: Lisbon, Portugal, 339–348, 1998.
216. Ehrhart P. Mds-Schwungrad-Energiespeicher Fir Mobile Anwendungen, in *Proceedings of Bmerien, Energierpeicher und Leisrungsspeicher Conference*. Essen: 1997.
217. Agboola AE. Development and model formulation of scalable carbon nanotube processes: Hipco and Comocat process models. Department of Chemical Engineering. Louisiana State University: Eunice, 2005.
218. Baughman RH, Zakhidov AA, de Heer WA. Carbon nanotubes—The route toward applications. *Science* 2002; 297: 787–792.
219. Li J, Ma W, Song L, et al. Superfast-response and ultrahigh-power-density electromechanical actuators based on hierarchal carbon nanotube electrodes and chitosan. *Nano Lett* 2011; 11: 4636–4641.
220. Baughman RH, Cui C, Zakhidov AA, et al. Carbon nanotube actuators. *Sci* 1999; 284: 1340–1344.
221. Baughman RH, Shacklette LW, Elsenbaumer RL, et al. Conducting polymer electromechanical actuators. In *Conjugated Polymeric Materials: Opportunities in Electronics, Optoelectronics, and Molecular Electronics*, J. L. Brédas, R. R. Chance, Editors. Kluwer, Dordrecht, Netherlands, 559–582, 1990.
222. Kaneto K, Kaneko M, Min Y, et al. Artificial muscle: Electromechanical actuators using polyaniline films. *Synth Met* 1995; 71: 2211–2212.
223. Osada Y, Okuzaki H, Hori H, et al. A polymer gel with electrically driven motility. *Nature* 1992; 355: 242–244.
224. Otero TF, Angulo E. Comparative kinetic studies of polypyrrole electrogeneration from acetonitrile solutions. *J Appl Electrochem* 1992; 22: 369–375.
225. Park H, Lieber C, Urban J, et al. Transition metal oxide nanowires, and devices incorporating them. U.S. Patent WO2003053851 A2, 2005.
226. Takeuchi I, Asaka K, Kiyohara K, et al. Electromechanical behavior of a fully plastic actuator based on dispersed nano-carbon/ionic-liquid-gel electrodes. *Carbon* 2009; 47: 1373–1380.
227. Rogers GW, Liu JZ. Graphene electromechanical actuation; Origins, optimization and applications. *MRS Proceed* 2012; 1407.
228. Liang J, Huang L, Li N, et al. Electromechanical actuator with controllable motion, fast response rate, and high-frequency resonance based on graphene and polydiacetylene. *ACS Nano* 2012; 6: 4508–4519.
229. Stoller MD, Park S, Zhu Y, et al. Graphene-based ultracapacitors. *Nano Lett* 2008; 8: 3498–3502.
230. Terrones M. Science and technology of the twenty-first century: Synthesis, properties and applications of carbon nanotubes. *Annu Rev Mater Res* 2003; 33: 419–509.
231. Peng H, Sun X, Cai F, et al. Electrochromatic carbon nanotube/polydiacetylene nanocomposite fibres. *Nat Nano* 2009; 4: 738–741.
232. Akle BJ, Bennett MD, Leo DJ. High-strain ionomeric–ionic liquid electroactive actuators. *Sensors Actuat A-Phys* 2006; 126: 173–181.
233. Liu S, Liu W, Liu Y, et al. Influence of imidazolium-based ionic liquids on the performance of ionic polymer conductor network composite actuators. *Polym Int* 2010; 59: 321–328.

234. Wilkes JS. A short history of ionic liquids-from molten salts to neoteric solvents. *Green Chem* 2002; 4: 73–80.
235. Dahiya S, Jain DK, Kumar A, et al. *Improvement of Adjustable Speed Drives (ASD's) Performance during Sag Conditions Using Ultracapacitors.* New Delhi, 2008.
236. Deswal SS, Dahiya R, Jain DK. Improved performance of an adjustable speed drives during voltage sag condition. *Int J Eng Sci Technol* 2010; 2: 2445–2455.
237. Jouanne AV, Enjeti P. ASD ride-through technology alternatives and development. Rep. EPRI TR-109903, Electric Power Research Institute: Palo Alto, CA, 1997.
238. Sullivan MJ, Vardell T, Johnson M. Power interruption costs to industrial and commercial consumers of electricity. *IEEE Trans Ind Appl* 1997; 33: 1448–1458.
239. Conrad L, Little K, Grigg C. Predicting and preventing problems associated with remote fault-clearing voltage dips. *IEEE Trans Ind Appl* 1991; 27: 167–172.
240. Wagner VE, Andreshak AA, Staniak JP. Power quality and factory automation. *Industry Applications Society Annual Meeting, Conference Record of the 1988 IEEE,* 1392: 1391–1396.
241. Zyl AV, Spee R. Short term energy storage for ASD ride-through. *Industry Applications Conference. Thirty-Third IAS Annual Meeting. The 1998 IEEE,* 1162: 1162–1167.
242. Schatz J. Flywheels improve power quality & reliability for southern company's customers. *Power Quality '9UPower Value '98 Proceedings.* 1998: 173–183.
243. Corley M, Locker J, Dutton S, et al. Ultracapacitor-based ride-through system for adjustable speed drives. *Power Electronics Specialists Conference. PESC 99. 30ᵗʰ Annual IEEE.* 1999: 26–31.
244. Hoffman D, Smith R. Ultracapacitors for the power quality market. *Power Systems, World '97:* Baltimore, 1997.
245. Enjeti PN, Duran-Gomez JL. Method and system for ride-through of an adjustable speed drive for voltage sags and short-term power interruption, The Texas A&M University Systems: United States, 1999.
246. Duran-Gomez JL, Enjeti PN, Jouanne AV. An approach to achieve ride-through of an adjustable-speed drive with flyback converter modules powered by super capacitors. *IEEE Trans Ind Appl* 2002; 38: 514–522.

Index